Computational Thinking for Life Scientist

Computational thinking is increasingly gainir
due to the unprecedented scale at which data
the cultural gap between the biological and computational ᴖᴄᴉᴇᴨᴄᴇ,
serves as an accessible introduction to computational concepts for students in the
life sciences. It focuses on teaching algorithmic and logical thinking, rather than
just the use of existing bioinformatics tools or programming. Topics are presented
from a biological point of view, to demonstrate how computational approaches
can be used to solve problems in biology such as biological image processing,
regulatory networks, and sequence analysis. The book contains a range of pedagogical features to aid understanding, including real-world examples, in-text
exercises, end-of-chapter problems, color-coded Python code, and "code
explained" boxes. User-friendly throughout, *Computational Thinking for Life
Scientists* promotes the thinking skills and self-efficacy required for any modern
biologist to adopt computational approaches in their research with confidence.

Benny Chor was a Professor in Computer Science (CS) at Tel-Aviv University,
Israel, and head of the School of Computer Science at Tel-Aviv University between
2018 and 2020. His research interests spanned over computational biology, crypt-
ography, and CS and math education, and he was renowned for his excellence in
teaching. Benny passed away in June 2021.

Amir Rubinstein is a lecturer in Computer Science at Tel-Aviv University, Israel.
His activity surrounds computer science education, and innovation and research
in teaching and learning CS. He has received numerous awards for outstanding
teaching.

"An excellent and very gentle introduction to bioinformatics for biologists. In contrast to books that focus on algorithms and ignore programming or focus on programming without explaining algorithms, this book is a perfect blend of both algorithms and programming!"

Pavel Pevzner, Ronald R. Taylor Chair and Distinguished Professor of Computer Science, University of California at San Diego

"The ability to extract quantitative information from data is an essential skill for the modern biologist. In order to maximize the benefit of programming, use of existing computational tools and effective collaboration with computational scientists, biologists must be able to 'think computationally' by gaining a more algorithmic and logical thinking. In their book, Benny Chor and Amir Rubinstein introduce fundamental computational concepts to life sciences students. Each chapter covers a distinct computational idea motivated by a concrete biological challenge. Questions embedded throughout each chapter and code examples provide hands-on practice. Similarly to the way in which chemistry is perceived as being essential to the biology curriculum, computational thinking should also be considered a part of the modern biologist's basic training. This excellent book is essential reading for undergraduate life sciences students."

Assaf Zaritsky, Ben-Gurion University of the Negev, Israel

Computational Thinking for Life Scientists

BENNY CHOR
Tel-Aviv University

AMIR RUBINSTEIN
Tel-Aviv University

CAMBRIDGE
UNIVERSITY PRESS

CAMBRIDGE
UNIVERSITY PRESS

University Printing House, Cambridge CB2 8BS, United Kingdom

One Liberty Plaza, 20th Floor, New York, NY 10006, USA

477 Williamstown Road, Port Melbourne, VIC 3207, Australia

314–321, 3rd Floor, Plot 3, Splendor Forum, Jasola District Centre, New Delhi – 110025, India

103 Penang Road, #05–06/07, Visioncrest Commercial, Singapore 238467

Cambridge University Press is part of the University of Cambridge.

It furthers the University's mission by disseminating knowledge in the pursuit of
education, learning, and research at the highest international levels of excellence.

www.cambridge.org
Information on this title: www.cambridge.org/9781107197244
DOI: 10.1017/9781108178327

© Benny Chor and Amir Rubinstein 2022

First published 2022

Printed in the United Kingdom by TJ Books Limited, Padstow Cornwall

A catalogue record for this publication is available from the British Library.

Library of Congress Cataloging-in-Publication Data
Names: Chor, Benny, author. | Rubinstein, Amir, 1977- author.
Title: Computational thinking for life scientists / Benny Chor, Tel-Aviv University, Amir Rubinstein, Tel-Aviv
    University.
Description: Cambridge, United Kingdom ; New York, NY : Cambridge University Press, 2022. | Includes
    bibliographical references and index.
Identifiers: LCCN 2021038567| ISBN 9781107197244 (hardback) | ISBN 9781316647592 (paperback)
Subjects: LCSH: Biology–Data processing. | Computational biology. | Computer logic.
Classification: LCC QH324.2 .C437 2022 | DDC 570.285–dc23
LC record available at https://lccn.loc.gov/2021038567

ISBN 978-1-107-19724-4 Hardback
ISBN 978-1-316-64759-2 Paperback

In loving memory of Benny Chor (1956–2021), my mentor, role model, and friend.

Contents

Introduction

What Is This Book About?

Modern biology has been undergoing a dramatic revolution in recent decades. Enormous amounts of data are produced at an unprecedented rate. These data come in various forms, such as gene and protein expression level data, DNA, RNA, and protein sequencing, high quality biological and medical images. One consequence of this "data explosion" is that computational methods are increasingly being used in life science research. Computational methods in this context are not the mere use of tools, but the integration of computational and algorithmic thinking to lab-experiment design; to data generation, integration, and analyses; and to modeling and simulation. It is becoming widely recognized that such thinking skills should be incorporated into the standard training of life scientists in the 21st century.

This book presents some fundamental ideas and notions from computer science in a biological context. Each chapter covers a biologically motivated challenge, and presents the relevant computational notions in a "user-friendly", non-formal manner. Each topic is accompanied by implementations in Python. This programming language has gained considerable popularity among biologists in recent years, as reflected by the proliferation of "Python for biologists" courses at university level around the world. This book also provides hands-on programming practice in the form of exercises throughout and at the end of each chapter.

Who Is This Book Aimed At?

This book is designed specifically for life scientists – students, teachers, researchers, and lab assistants – who wish to acquaint themselves with the computational "culture," including (but not limited to) basic programming skills.

This book assumes basic knowledge in biology. Any introductory course in biology should suffice. Additionally, no background in computer science or programming is assumed, as basic programming is introduced in chapter 1.

We note that this book does not cover any specific bioinformatics tool handling content. Such tools are covered in dedicated guides. Instead, our main goal is to develop fundamental computational thinking skills, which are still absent from many biology education programs.

What Skills Should I Expect to Gain from This Book?

You should be able to:

- identify problems for which a manual solution is not viable, or reaching such a solution is very tedious and error prone, yet such problems are amenable to a computational solution
- implement simple solutions to some biologically motivated problems using the programming language Python
- understand various fundamental notions in computer science and their applicability to the life sciences
- more effectively communicate and collaborate with experienced programmers, computer scientists, and computational biologists

What Topics Are Covered?

The book is divided into five parts. Part I starts with an introduction to programming in Python (chapter 1), which includes most of the programming background needed for later chapters. The important notion of computational complexity as a measure for the efficiency of algorithms is then presented (chapter 2). Part II deals with problems that involve biological sequences (e.g., DNA, RNA, proteins), and familiarizes the reader with two related notions in computing: hashing (chapter 3) and regular expressions (chapter 4). In Part III, we focus on biological networks, and introduce some basic notions in the important field of graph theory (chapter 5), provide an example for a graph-based algorithm – breadth-first-search (chapter 6), and showcase a simulation methodology for regulatory networks (chapter 7). Part IV deals with the representation and processing of biological images (chapters 8 and 9). Finally, part V exposes some fundamental limitations of computing – such as the P vs. NP open problem and the inability to solve many well-defined problems – and offers some deeper concepts from the theory of computer science.

Do You Have a Website?

Yes, at www.cs.tau.ac.il/~amirr/book.

On the book's website, you will find all the Python code demonstrated in the book, as well as some solutions to exercises, notes, and additional recommended resources.

Notes for the Reader

Practical experience is essential for the acquisition of programming skills. Therefore, we advise having a Python session (more details below) open alongside reading the book. We encourage you to download the code from our website, and "play" with it while reading: for example, execute it with various inputs, change something and inspect the effect, and try to solve a problem in a different way. You may find yourself struggling with the code to make it work properly, which can be frustrating. However, we think this is not a waste of time – it is a necessary trial-and-error-based learning process.

Each chapter includes various exercises, of two types: short questions embedded in the chapter, and "challenge yourself" problems at the end of the chapter that explore specific aspects in more depth. Together these should provide you with "hands-on" practice, as well as encourage and develop your computational thinking.

The parts of this book are mostly independent. Therefore, you may choose to focus on specific topics, without having to read all the previous chapters (with the exception of chapter 1 that provides the programming background for the rest of the book and chapter 2 that introduces the notion of complexity).

Notes for the Teacher

Biology teachers and lecturers who are interested in incorporating some computational content into their courses may find some ideas in this book. From our experience, the whole content of this book fits within a 13-week semester at university. If you wish to devote only a small number of hours for that in your course, just pick a single part in the book (or even chapter) that is most appropriate to your audience in terms of background and pedagogical interests. We kept "backwards dependencies" minimal, and you should be able to pick any part without having to cover previous ones for background.

The problems presented at the end of each chapter ("challenge yourself") may be used as a basis for small projects, which explore the topic even deeper.

Python Preliminaries

Computer programs are sequences of commands that follow a very strict and formal set of rules (syntax). A specific syntax defines what we call a programming language, such as Python, Java, or C. Programming languages are also termed "high-level" languages, because their execution involves, as a first step, their transformation into a language that a computer can understand and execute, termed "low-level" or machine language (this is basically a sequence of 0s and 1s, corresponding to two separable physical states, e.g., voltage). In Python, this transformation is done in a process called "interpretation" (in languages such as Java or C the process is somewhat different and is termed "compilation"). Interpretation is done by special software called an *interpreter*. There are several interpreters for Python which you can download. Possibly the simplest one is called IDLE, the standard Python interpreter. On the course website, you can find links and download instructions.

Executing commands in Python can be done in two modes: *interactive* mode, in which a single command is written, executed, and immediately returns the result to the user, and then the user is prompted for the next command, and so on; and *script* mode, in which a program is written in an editor, stored in a file with a .py extension (e.g., my_program.py), and then executed as a batch of commands one by one. The former mode is convenient for the execution of short and specific computations, while the latter is suitable for writing programs that we intend to execute multiple times.

In this book, we will use the following format conventions regarding these two modes. Commands in interactive mode are presented in a shell-like format, in which lines start with a prompt sign >>>, and their results appear in the next line or lines:

```
>>> print(3+4)
7
```

Programs written in script mode will be presented in an editor-like format, with lines numbered so we can refer to specific lines in the text:

```
1 def max(x, y):
2     if x>y:
3         return x
4     else:
5         return y
```

The execution of such scripts will be presented as interactive mode executions:

```
>>> max(8, 3)
8
>>> max(2, 4+1)
5
```

We follow the default syntactic color highlighting of IDLE.

We hope this book will expose to you the beauty of the computational "culture," by introducing a diverse range of computational concepts that are applicable to different biological settings.

Finally, we wish to deeply thank Smadar Gazit for her support in writing and organizing this book, always with a smile and good will !

Cheers,
Benny and Amir

Part I: Programming in Python

– "First things first."

In this part of the book, we will introduce the basics of the programming language Python (chapter 1), and learn how to use it efficiently, taking into consideration a program's running time and memory allocation requirements (chapter 2). Later on, Python programs will accompany every topic presented throughout this book.

The syntax and usage of Python is presented in chapter 1 mostly through examples. We recommend executing these examples, and playing with them a bit. We believe this is an efficient way to learn a new programming language for a novice, rather than thoroughly presenting the syntax in a formal and rigorous manner. While the examples presented are not comprehensive, they do provide the necessary ground for a useful (and hopefully enjoyable) reading of the rest of this book.

If the notion of computational complexity is new to you, chapter 2 will provide some familiarity with this fundamental concept. Again, the topic will be presented rather informally, focusing on the intuition rather than the precise definitions. At the end of that chapter, you should be able to appreciate how important it is to devise efficient algorithms. For example, sometimes having an inefficient algorithm at hand is practically equivalent to having no algorithm at all. Efficiency comes in two flavors – time complexity, relating to an algorithm's running time, and memory complexity, relating to memory allocation requirements of an algorithm. We will mention both, as both are relevant in computational biology, where we deal with very large inputs, such as a whole genome or a large network of interactions.

1 Crash Introduction to Python

This chapter kicks off with an example for a simple program in Python. Through this example, we will delve right away into the syntax of this popular programming language. There will be quite a few technical details in this example, which are probably new to anyone with no prior Python experience. The rest of the chapter will then walk the reader briefly yet methodologically through the basic Pythonic syntax. By the end of this chapter, you will be able to write simple programs in Python to solve various problems of a computational nature.

We recommend reading this chapter, and in fact the entire book, alongside with a Python working environment. The code you will see also appears on the book's website, and running it in parallel to reading the material will most certainly improve your understanding, and allow you to explore beyond what we present. Solving the exercises is highly recommended before moving on.

1.1 A Crash Example

Please welcome our first program in Python for this book!

```
1  def GC_content(dna_string):
2      ''' calculate %GC in dna_string (uppercase ACTG) '''
3      GC = 0
4      for nuc in dna_string:     # iterate over dna_string
5          if nuc == "C" or nuc == "G":
6              GC += 1            # equal to GC = GC+1
7      return 100*GC/len(dna_string)
```

Code explained

This code defines a function called GC_content. This function receives a single parameter, whose name is dna_string. The parameter dna_string is expected to be a string that consists of the characters A, T, C, and G only. However, this function does not verify this. Furthermore, in Python variable types are not specified, and we will get back to this point soon. Note the colon at the end of the "title" row.

The body of the function appears next. In Python, indentation defines scopes: we know that lines 2–7 are the definition of the function's body because all these lines are indented to the right. Similarly, code inside loops or conditional statements is also indented, as we will discuss very soon.

Line 2 is a documentation string (also termed "docstring"): this is a string that appears between triple quotes and merely describes the function. It has no effect in terms of the function's operation. Line 3 defines a variable called GC, and initializes it to 0. In line 4, there is a loop that iterates over the elements of the input variable dna_string. This line should be understood as "use a variable named nuc and assign it to the characters of `dna_string` one by one." We chose the name `nuc` because each of these characters represents a single nucleotide. The # character denotes the beginning of a programmer's comment (comments have no effect on the program's functionality). In line 5, we use an if-statement (termed conditional statement) to check whether `nuc` is equal to `"C"` (Cytosine) or to `"G"` (Guanine). The equality operator in Python is the equals sign twice, `==`, and the two sub-conditions are combined with a logical `or` relation (termed logical operator). If the condition is true, we increase the variable `GC` by `1`, using the `+=` operator: `GC += 1` is merely a shortcut for `GC = GC+1`. By the end of the loop in lines 4–6, the variable `GC` stores the number of characters in `dna_string` that are `"C"` or `"G"`. Finally, in the last line, the function returns the percentage of `"C"` and `"G"` in the input string to the calling environment.

Once we have written this function, we can use it by calling it:

```
>>> GC_content("AAA")
0.0
>>> GC_content("CCG")
100.0
>>> GC_content("TTAGACCAGTAGAAGTAC")
38.888888888888886
```

The programing notions presented in this example are: function definition and function call, variable, string, loop, if statement, operator, and return value. Let's go over these in more detail now.

1.2 Variables, Basic Types, and Operators

Computer programs use **variables** to store data. Data values in Python and the variables storing these data can have different types. Table 1.1 summarizes several common types in Python.

Numbers in Python come in two types: **integers** of type `int`, for whole numbers, and numbers with a **decimal point** of type `float` (such numbers are represented in the computer's memory in the floating point method, which we will not explain here, thus the type name).

Type `str` is used to store **strings**. A string is a sequence of characters, appearing between quotation marks. Note that strings in Python may be written between double or single quotation marks ("..." or '...'). Triple-double ("""...""") or triple-single ('''...''') quotation marks are also used in Python in two cases: as documentation strings at the beginning of

Table 1.1. *Variable types in Python*

Type/class	Meaning	Examples for values of this type
int	integer	`0, 1, -1, 2, -2, 3, -3...`
float	rational number	`3.14     -0.001     8.0`
str	string, sequence of characters	`"Amir"   "!!!"   "45"   "ATTCG"`
list	sequence of elements	`[1, 2, 3]   [55, "hello", 55, -7 ]`
bool	Boolean	`True / False`

functions (as we saw in the earlier example), and for strings that span over more than one line (we will see this later). In any case, a string is started and terminated by matching'terminators.

Type `list` is used to store, well, **lists**, which are sequences of elements of any type. Elements in a list are separated by commas, and enclosed within square brackets [ ].

In Python, types are also called **classes**. In fact, every object in Python belongs to a class, which defines its functionality. Python has a built-in function named `type` that returns the class of a value or variable:

```
>>> x = 5
>>> type(x)
<class 'int'>
>>> y = 0.001
>>> type(y)
<class 'float'>
>>> type(True)
<class 'bool'>
>>> s1 = "drosophila"
>>> type(s1)
<class 'str'>
>>> type("45")
<class 'str'>
>>> lst = [1, 2, 3, "hello", 0.001]
>>> type(lst)
<class 'list'>
```

In Python, unlike some other programming languages, such as C and Java, there is no explicit declaration of the type of a variable. This is because variables in Python have dynamic types, which may change whenever a new value is assigned to the variable:

```
>>> x = 5
>>> type(x)
<class 'int'>
>>> x = "drosophila"
>>> type(x)
<class 'str'>
```

Each type in Python supports specific **operators** that can be applied to it. For example, numeric types (`int` or `float`) support the arithmetical operators in Table 1.2.

As you can see, Python[1] has two division operators: one is for "true" division and the other rounds the result down and yields an integer (this is why it is also called integer division):

```
>>> 8/5
1.6
>>> 8//5
1
```

---

[1] In fact Python version 3, but not the older version 2.

Table 1.2. *Numeric operators*

Operator	Symbol	Example
Addition	+	x+y
Subtraction	−	x−y
Negation	−	−x
Multiplication	*	x* y
Division	/	x/y
Floor division	//	x//y (this is x/y rounded down)
Exponentiation	**	x** y (this is x^y)
Modulo	%	x%y (the remainder of x divided by y)

Table 1.3. *Comparison operators*

Operator	Symbol	Example
Equal	==	x == y
Unequal	!=	x != y
Less than	<	x < y
Less or equal	<=	x <= y
Greater than	>	x > y
Greater or equal	>=	x >= y

Python defines an **order of precedence** for operators. For example, the expression 3*4+5 will result in 17, since multiplication precedes addition, just as you have learned in elementary school (exponentiation precedes multiplication and division, which in turn precede addition and subtraction). Parentheses are used to impose a different order of evaluation:

```
>>> 3*4+5
17
>>> 3*(4+5)
27
```

We do not find much point in memorizing the precedence order defined by Python, thus we do not present it here. Instead, we recommend using parentheses when in doubt. We will see some operators for strings and lists later in this chapter.

1.3 Comparison, Logical, and Assignment Operators

Python supports several **comparison operators**, as summarized in Table 1.3. The result of a comparison is either `True` or `False`, which are termed **Boolean** values (after the English mathematician George Boole). Expressions of Python's type `bool` are termed Boolean expression, or logical expressions.

```
>>> x = 3
>>> y = 4
>>> x == y
False
>>> x == y-1
True
>>> x != 3
False
>>> x < y
True
>>> s1 = "drosophila"        <= this is an assignment
>>> s1 == "Drosophila"       <= this is a comparison
False
>>> [1,2,3] == [2-1,2,2+1]
True
```

Exercise 1

Not every two types are orderable, that is, can be compared by the $<, <=, >, >=$ operators. Try for example:

```
>>> 5 < "5"
```

And observe the resulting error message. However, strings in Python are orderable. You may wonder how order is defined for strings. One could imagine that the order is defined merely by length, but this can be ruled out by a simple check:

```
>>> "cat" < "astronaut"
False
```

Try to figure out how the order relation is defined for strings, by inspecting the following examples, and running additional ones that will support your hypothesis.

```
>>> "cat" < "dog"
True
>>> "cat" < "bee"
False
>>> "cat" < "catalogue"
True
```

Boolean expressions can be combined to form more complex expressions, using the **logical operators** "and," "or," and "not". Here is a short reminder (A, B in Table 1.4 are Boolean expressions):

Table 1.4. *Boolean operators*

Operator	Usage	When this expression is True
and	A and B	both A and B are True
or	A or B	at least one of A and B is True
not	not A	A is False

```
>>> x = 3
>>> y = 4
>>> x == 3 and x == y
False
>>> x == y or x == y-1
True
>>> not (x == y or x == y-1)
False
```

The negation operator may be convenient when negating a non-trivial condition. In the last example, we could easily replace

```
not (x == y or x == y-1)
```

by

```
x != y and x != y-1,
```

using simple logical rules. However, when the condition is more complex, negation with the `not` operator may be more convenient.

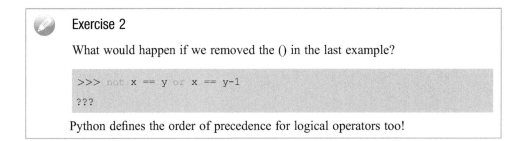

Exercise 2

What would happen if we removed the () in the last example?

```
>>> not x == y or x == y-1
???
```

Python defines the order of precedence for logical operators too!

In many of the previous examples, we used an **assignment operator**, denoted =. There are several additional operators that involve assignment, which actually combine assignment with modification of a variable. In the `GC_content` example, we already saw this: `GC += 1` is a shortening for `GC = GC+1`. Other assignments of this kind are `-=`, `*=`, `/=`, `//=`, `**=` and even `%=`. Here are several examples (note especially the last one on strings):

```
>>> x = 3
>>> y = 4
>>> x += 1
>>> x
4
>>> x *= 2
>>> x
8
>>> x -= 2*y      <= stands for x = x-2*y
>>> x
0
>>> s = "AAA"
>>> s += "GGG"
>>> s
"AAAGGG"
```

1.4 Type Conversions

Sometimes we need to perform computations that mix more than one type. For example:

```
>>> x = 5 + 3.2
```

Exercise 3

(a) What do you think is the result of the above computation? What is its type?

(b) Suppose you had to vote for the resulting type of the following computation:

```
>>> x = 0.8 + 1.2
>>> type(x)
???
```

The result is obviously the number 2, but is it 2 of type `int` or 2.0 of type `float`? What would you vote for? Why? Check if this is the case in Python as well.

In some cases, there is no right or wrong answer, but rather a decision of the designers of the programming language, according to what they find most appropriate, taking into account various considerations, such as consistency, ease of use, or error prevention.

When we mix `float` and `int`, Python first converts the `int` value into `float` (e.g., 5 is converted to its `float` parallel 5.0), and then the expression is computed. This makes sense because often we need to make computations that involve integers and real numbers. Such conversions are called **automatic type casting**.

Exercise 4

Do you think Python would allow adding up a number and a string, involving an automatic casting from `int` to `str` or vice versa? Try for example:

```
>>> x = "5" + 3.2
???
```

Some type mixing is illegal and yields an error. This is the case when the language designers think that a certain operation is more likely to be a bug rather than the intension of the programmer. An error will prevent such commands from going unnoticed.

In addition to the automatic casting done by Python when mixing values of appropriate different types, Python also allows **explicit casting**. Such type casting is forced by the programmer, and is an important feature as we will see in many examples in this book. However, it should be used with caution, as some conversions do not make sense. The explicit conversion in the next examples is done through the functions `int()`, `float()`, `str()`, and `list()`.

```
>>> x = 5.2
>>> int(x)
5

>>> y = 8
>>> float(y)
8.0

>>> str(56)
'56'

>>> int("34")
34

>>> int("hello")
Traceback (most recent call last):
  File "<pyshell#0>", line 1, in <module>
    int("hello")
ValueError: invalid literal for int() with base 10: 'hello'
```

As you can see, converting from string to integer may yield an error – not everything can be converted to a number. When we ask Python to execute an illegal command, it informs us that there is some problem with the command. Sometimes, the error message is very informative, and sometimes it may look a bit cryptic and more experience will be required to understand it. In this example, the error message may look uninformative, but it merely says that Python tried to interpret the string "hello" as a decimal number, and failed.

Exercise 5

Does the direct conversion from float to int round a number to the closest integer or always round the number down? Is the result of the following 5 or 6?

```
>>> x = 5.8
>>> int(x)
???
```

Conversion from a string to a list creates a list that contains each character in the string as a separate element:

```
>>> list("TATAAA")
['T', 'A', 'T', 'A', 'A', 'A']
```

The other direction, however, is a bit trickier. Let's try:

```
>>> str(['T', 'A', 'T', 'A', 'A', 'A'])
"['T', 'A', 'T', 'A', 'A', 'A']"     <= not what we'd want!
```

What we have here is not a string with the characters contained in the list, as we might expect. The commas between the elements, the quotations marks around each character, as well as the [ ] surrounding the list all became characters in the new string! Although this is a bit early in terms of Python features we have learned, we show below how this conversion should be done appropriately. We note that in Section 1.6.3 we explain the syntax shown here in more detail.

```
>>> "".join(['T', 'A', 'T', 'A', 'A', 'A'])
"TATAAA"
```

1.5 Strings and Lists as Sequences

Like the numerical operators we saw earlier, strings and lists each have their own operators. Since both types are ordered sequences of elements, they have several common operators. Please see two useful operators of strings and lists in Table 1.5.

Here is another example:

```
>>> positive_aa = ["Lysine", "Arginine"]
>>> negative_aa = ["Aspartic acid", "Glutamic acid"]
>>> charged = positive_aa + negative_aa
>>> charged
["Lysine", "Arginine", "Aspartic acid", "Glutamic acid"]
```

Table 1.5. *Sequence operations*

Operator	Symbol	Example for strings	Example for lists
Concatenation	+	>>> "droso" + "phila" "drosophila"	>>> [1,2,3] + [4,5] [1,2,3,4,5]
Duplication	*	>>> "UAG" * 8 "UAGUAGUAGUAGUAGUAGUAGUAG"	>>> ["a",4] * 3 ["a",4,"a",4,"a",4]

Note that assignment operators =+ and *= also work for strings and lists:

```
>>> dna = ""              <= an empty string
>>> dna += "UAG"
>>> dna
'UAG'
>>> dna += "CC"
>>> dna
'UAGCC'

>>> charged = []          <= an empty list
>>> charged += ["Lysine", "Arginine"]
>>> charged += ["Aspartic acid", "Glutamic acid"]
>>> charged
["Lysine", "Arginine", "Aspartic acid", "Glutamic acid"]
```

When working with strings and lists, **indexing** is used to access characters or elements in specific positions (indices). We use [i] to probe the ith element. The first position is accessed with the index 0, the second one with the index 1, etc. So, legal indices are integers between 0 and the length of the string/list minus 1.

```
>>> organism = "drosophila"
>>> organism[0]
'd'
>>> organism[1]
'r'
>>> organism[5]
'p'
>>> organism[9]
'a'
>>> organism[10]
Traceback (most recent call last):
  File "<pyshell#6>", line 1, in <module>
    organism[10]
IndexError: string index out of range
```

In the last error message, Pythons tells us we tried to access a string with an illegal index, since the last legal index of a string of length 10 is 9.

Exercise 6

What do you think will the result of these commands be?

```
>>> positive_aa = ["Lysine", "Arginine"]
>>> positive_aa[0][0]
???
>>> positive_aa[1][3]
???
```

The last exercise shows an example for indexing of "higher dimensions," that is, we access the *i*th element of a sequence (string or list), and if this element is a sequence itself, we can index into it as well. This will be very useful later in the book, when we work with lists of strings, or lists containing inner lists. The latter will be used to store tables or matrices. For example, here is a 3×3 matrix, represented as a list of lists:

```
>>> mat = [ [1,2,3], [4,5,6], [7,8,9] ]
>>> mat[1][2]
6
```

Checking the **length** of a sequence in Python is easy, using the built-in function `len()`:

```
>>> organism = "drosophila"
>>> len(organism)
10
>>> organism[len(organism)-1]
'a'
>>> organism[len(organism)]
Traceback (most recent call last):
  File "<pyshell#10>", line 1, in <module>
    organism[len(organism)]
IndexError: string index out of range
>>> positive_aa = ["Lysine", "Arginine"]
>>> len(positive_aa * 2)
4
```

Exercise 7

What is the output of the following length tests?

```
>>> mat = [ [1,2], [3,4], [5,6] ]
>>> len(mat)
???
>>> len(mat[0])
???
```

Python allows **slicing** a sub-sequence from a given sequence, which is a very convenient feature. To slice a string or list, the start and the end indices are provided between [ ] separated by a colon. The resulting slice contains all the elements within the range, not including the end position:

```
>>> organism = "drosophila"
>>> organism[0:4]
'dros'
>>> organism[0:1]
'd'
>>> organism[0:len(organism)]
'drosophila'
```

We can also provide a third integer, to denote the "leap" within the range of indices (otherwise the default leap is 1):

```
>>> organism[0:len(organism):2]
'doohl'
>>> organism[9:3:-1]  <= leap of one position backwards
'alihpo'
```

Exercise 8

When we omit the start and/or end indices, Python uses default values for them. What are these default values? Check the output, for example, in the following cases:

```
>>> "drosophila"[3::2]
???
>>> "drosophila"[::-1]
???
```

(The last command is worth remembering, that's an easy way to **reverse** sequences in Python.)

The last useful functionality of strings and lists we present in this section is **membership checking**, using Python's operator `in`.

```
>>> organism = "drosophila"
>>> "p" in organism
True
>>> "phila" in organism
True
>>> "c" in organism
False
>>> 1 in [1,2,3]
True
>>> "1" in [1,2,3]
False
```

Finally, we remark that strings and lists have many differences too. One such fundamental difference is that lists are **mutable**, while strings are **immutable**. This simply means that we can change elements in a list, but we cannot change characters in a string.

```
>>> lst = ["T", "A", "A", "T", "A"]
>>> lst[1] = "G"
>>> lst
["T", "G", "A", "T", "A"]

>>> dna = "TAATA"
>>> dna[1] = "G"
Traceback (most recent call last):
  File "<pyshell#19>", line 1, in <module>
    dna[1] = "G"
TypeError: 'str' object does not support item assignment
```

One way to bypass this limitation is by converting a string to a list, changing (mutating) its elements as needed and then converting them back to a string.

```
>>> dna = "TAATA"
>>> dna_list = list(dna)
>>> dna_list[1] = "G"
>>> dna = "".join(dna_list)
>>> dna
'TGATA'
```

This may look a bit cumbersome, but it does the job.

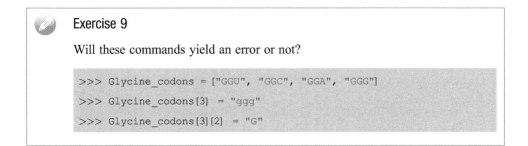

Exercise 9

Will these commands yield an error or not?

```
>>> Glycine_codons = ["GGU", "GGC", "GGA", "GGG"]
>>> Glycine_codons[3] = "ggg"
>>> Glycine_codons[3][2] = "G"
```

In the next section on functions, we will see more differences between strings and lists: each type supports different functions.

1.6 Functions

1.6.1 Python Built-In Functions

In the previous sections, we saw several commands in Python that involved **functions**. For example, we have encountered `type(...)`, `int(...)`, `str(...)`, `len(...)`, etc. These were all examples for **built-in functions** in Python. This means that these functions are part of the language, and are ready to use whenever we write a program in Python.

Python has many more built-in functions, for example `min`, `max`, and `sum`. Given a list of numbers, these functions return the minimum, maximum, and sum of the list:

```
>>> min([5,2,-4,7,8])
-4
>>> max([5,2,-4,7,8])
8
>>> sum([5,2,-4,7,8])
18
```

You may want to check what happens when these functions are given a list that contains not only numbers, or an input that is not a list at all.

> **Exercise 10**
>
> Try calling the above functions on a list containing only strings as elements. Is this a legal input for each of them?
>
> ```
> >>> min (["cat", "dog", "bee"])
> ???
> >>> sum (["cat", "dog", "bee"])
> ???
> ```

Another useful Python built-in function is `sorted`:

```
>>> sorted([5,2,-4,7,8])
[-4,2,5,7,8]
```

> **Exercise 11**
>
> Does Python's function `sorted` change the list it received as input or does it create a sorted copy of the list? Check this:
>
> ```
> >>> L = [5,2,-4,7,8]
> >>> sorted(L)
> [-4,2,5,7,8]
> >>> L
> ???
> ```

The command `help(__builtins__)` will present all the built-in functions (and other objects, such as classes) that exist in Python. However, in this book we will be using a very small portion of those, and present them when needed.

1.6.2 User-Defined Functions

Writing one's own functions in Python is really necessary for being able to design new programs. This chapter kicked-off with an example of a function that computes the GC content of a DNA sequence given as input. Here is another example for a simple function that takes a number representing the temperature in Fahrenheit and converts it to Celsius (the formula is $°C = (°F − 32)/1.8$).

```
1 def fahr2cel(fahr_temp):
2     ''' input is temperature in Fahrenheit, output in Celsius '''
3     cel_temp = (fahr_temp - 32) / 1.8
4     return cel_temp
```

```
>>> fahr2cel(86)
30.0
>>> fahr2cel(104.6)
40.33333333333333
```

The general structure for a function's definition is this:

```
def func_name(parameters):
<-tab->    function body
       .
       .
       .
       return value
```

The body of the function includes statements, which are indented by a tab (or a fixed number of spaces). We will see many more user-defined functions in the rest of this chapter, and in the rest of this book.

1.6.3 Class Methods

In addition to the built-in and user-defined function, additional sources for functions are **class methods**. These are functions that belong to specific classes in Python. For example, there are methods that belong to the class of strings. Here are some examples:

```
>>> dna = "ATTGCGGGCTTG"
>>> dna.count("C")
2
>>> dna.count("G")
5
>>> dna.lower()
'attgcgggcttg'
>>> dna.replace("G", "g")
'ATTgCgggCTTg'
>>> dna.find("G")       <= the first index where 'G' is found in dna
3
>>> dna.find("g")       <= if not found
-1                      <= then -1 returned
>>> dna
'ATTGCGGGCTTG'          <= remained unchanged
```

Code explained

The variable `dna`, which belongs to the `str` class, can invoke any method from class `str`. This it does using ".". (dot) between the object name and the required method:

```
object.method(parameters)
```

The method `count` receives a string parameter and returns the number of times the parameter appears in the activating string. The method `lower` receives no parameters, and returns a copy of the activating string converted to lower case (without changing it). The method `replace` takes two parameters and replaces every occurrence (in the activating string) of the first one with the second one. The method `find` returns the first occurrence of the input character in the activating string. These are only a few of the methods in class `str`.

Exercise 12

The method `count` can get as input more than a single character. For example:

```
>>> "ATTGCGGGCTTG".count("GC")
2
```

What do you think will be the output in the next example? What can you conclude about how the method `count` treats overlaps in the searched pattern occurrences?

```
>>> "TATAT".count("TAT")
???
```

Similarly, the list class also has its own methods. Here are some useful ones:

```
>>> L = [3,1,1,2,1,4]
>>> L.count(1)
3
>>> L.append(5)          <= add element at the end of the list
>>> L
[3, 1, 1, 2, 1, 4, 5]
>>> L.insert(3,100)      <= insert 100 at position 3
>>> L
[3, 1, 1, 100, 2, 1, 4]
>>> L.pop(3)             <= remove element in position 3
100
>>> L
[3, 1, 1, 2, 1, 4]
>>> L.remove(2)          <= remove the value 2 from the list
>>> L
[3, 1, 1, 1, 4]
>>> L.reverse()          <= reverse the order of the list
>>> L
[4, 1, 1, 1, 3]
>>> L.sort()             <= sort the list
>>> L
[1, 1, 1, 3, 4]
```

🗨 **Code explained**

Class `list` also has a method called `count` that works similarly to `str`'s `count`. The only difference is that the parameters are not limited to a string type. `append` adds an element at the end of the list. The method `insert` takes two parameters: the position at which to insert a new element, and the new element to insert. The method `pop` is used to delete an element by its index in the list, while `remove` deletes an element with a specified value; `reverse` and `sort` both change the order of the elements in the activating list in the way reflected by their names. Note that Python's built-in function `sorted` is different from `list`'s `sort`, in that the former receives the list *as a parameter* and creates a sorted copy of it.

✏ **Exercise 13**

The method `remove` in the class `list` removes an element from the list whose value is given as a parameter. But what happens when there is more than one such element?

```
>>> L = [1,0,0,1,2,1,4]
>>> L.remove(1)
>>> L
???
```

Knowing which methods exist in each class will definitely pay off, as it saves a lot of time and effort writing many lines of code from scratch. We refer the curious readers to Python's documentation, to your favorite search engine, or to the commands `help(str)` and `help(list)` in order to explore the various class methods available. However, whenever we use a method for the first time in this book, we will explain it.

Functions can be used inside the body of other functions. For example, look at the following user-defined function `GC_content2`, which has the same functionality as the old `GC_content` from the beginning of this chapter. This function uses the method count of class string.

```
1 def GC_content2(dna_string):
2     ''' calculate %GC in dna_string (uppercase ACTG) '''
3     C = dna_string.count("C")
4     G = dna_string.count("G")
5     return 100 * (C+G)/len(dna_string)
```

✏ **Exercise 14**

What would happen if we called `GC_content2` with an input parameter that contains the DNA sequence in lowercase? Try this for example:

```
>>> dna = 'ttagaccagtagaagtac'
>>> GC_content2(dna)
???
```

Suggest a way to overcome this so that the function will be case insensitive.

1.6.4 Return Value(s)

A function's execution is terminated when a `return` command is encountered. The value that appears after the `return` word is propagated back to the calling environment, and can be assigned to a variable for further use:

```
>>> result = GC_content("TTAGACCAGTAGAAGTAC")
>>> print(result)
38.888888888888886
```

Sometimes we write functions but do not need them to return anything. In such a case, we can write `return None` (or simply `return`) at the end of the function, or completely omit the return command. In any of these cases, the function will return the special value `None`, which is the Pythonic way to say "nothing" (which is, as you know, different from not saying anything!).

```
1 def modify(list, i, value):
2     list[i] = value
3     return None
```

```
>>> L = [1,2,3]
>>> res = modify(L, 0, 99)
>>> print(L)
[99,2,3]
>>> print(res)
None
```

Exercise 15

Look at the following function F and its usage:

```
def F(x):
    x += 1

>>> res = F(7)
>>> res
???
```

What is the result? Can you explain this? Will the answer change if F is changed to the following?

```
def F(x):
    x += 1
    return x
```

1.6.5 Default Parameters

Parameters of functions can have **default values**. For example:

```
1 def F(a=10, b=20):
2     return a*b
```

The default value for the first parameter is 10, and for the second the value is 20. This means that when we call the function, we may provide less than two parameters, in which case the default values will be assigned to the missing ones:

```
>>> F()
200
>>> F(a=7)
140
>>> F(b=13)
130
>>> F(a=4, b=6)
24
>>> F(4,6)
24
```

Note that we can still call the function normally as we used to, as in the last command above.

1.7 Conditional Statements

The function GC_content from the beginning of the chapter used a **conditional statement**, starting with 'if'. Such a statement splits the program's flow into two or more possible paths. Here is another example: suppose we want to know the physical state of some chemical substance. We get as input the melting (Tm) and vaporization (Tv) temperatures of the substance, and the room temperature, T:

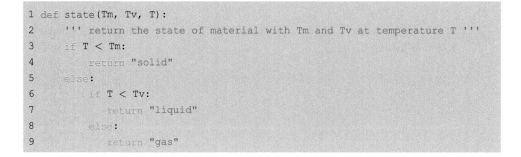

```
1 def state(Tm, Tv, T):
2     ''' return the state of material with Tm and Tv at temperature T '''
3     if T < Tm:
4         return "solid"
5     else:
6         if T < Tv:
7             return "liquid"
8         else:
9             return "gas"
```

Code explained

The function checks the value of the condition T < Tm. In case it is True, the function terminates and returns the string "solid". Otherwise, it moves on and checks whether T < Tv, in which case "liquid" is returned. If both conditions were False, "gas" is returned. Another way to write an equivalent conditional statement would be:

```
if T < Tm:
    return "solid"
elif T < Tv:
    return "liquid"
else:
    return "gas"
```

note that elif is a shortcut for "else if".

The general structure for conditional statements is this:

```
if condition:
    <-tab-> statement
        ...
elif condition:        # 0 or more
    <-tab-> statement
        ...
else:                  # optional
    <-tab-> statement
        ...
```

Often the conditions in a program are composed of several "smaller" conditions, combined by logical operators. In the GC_content example, we checked whether the current nucleotide is either cytosine or guanine, and used the logical operator or.

1.8 Iteration and Loops

1.8.1 "for" Loops

Recall the **loop** in the first example of this chapter (`GC_content`). Such loops are usually used in Python to iterate over a collection of elements, for example strings or lists. Here is an additional example:

```
1 def basic_aa_perc(prot):
2     ''' return percentage of basic amino acids in protein '''
3     basic = 0
4     basic_aa_list = ["K", "H", "R"]
5     for aa in prot:
6         if aa in basic_aa_list:
7             basic += 1
8     return 100*basic/len(prot)
```

> **Code explained**
>
> The function computes the percentage of the amino acids lysine, histidine, and arginine in the protein input sequence. In line 5, the variable `aa` (short for amino acid) is implicitly defined, and then assigned characters from `prot` one by one in each iteration. For any such `aa`, if it is either `"K"`, `"H"` or `"R"` the counter is increased.

The general structure of "for" loops is the following (The collections we have seen so far were strings and lists. We will see later that "for" loops are also applicable to other collections.):

```
for var in collection:
<-tab-> statement
        ...
statement          # out of loop
```

> **Exercise 16**
>
> The function, `basic_aa_perc` iterates over the input protein `prot`. We could instead go over the list of basic amino acids. Complete the missing line in the following functions, using class `str`'s method count:
>
> ```
> def basic_aa_perc(prot):
>     basic = 0
>     basic_aa_list = ["K", "H", "R"]
>     for basic_aa in basic_aa_list:
>             complete this line
>     return 100*basic/len(prot)
> ```
>
> Do you expect one of these versions to run faster? We will see later on how to answer such questions, and even measure and compare actual running times.

1.8.2 "while" Loops

There is another type of loops in Python – "while" loops. As a first example, let us re-write the GC_content function:

```
1 def GC_content3(dna_string):
2     ''' calculate %GC in dna string (uppercase ACTG) '''
3     GC = 0
4     i = 0
5     while i < len(dna_string):
6         if dna_string[i] == "C" or dna_string[i] == "G":
7             GC += 1
8         i += 1
9     return 100*GC/len(dna_string)
```

Code explained

The variable i is used as the running index. It is initialized to 0, used to access specific elements in the input dna_string in line 6, and increased by 1 each iteration in line 8. The loop repeats as long as i is smaller than the length of dna_string, as specified in line 5. Note that if we forgot to increase i in line 8, we would get an infinite loop that would run (theoretically) forever!

The general structure of "while" loops is the following:

```
while    condition:
<-tab-> statement
        ...
statement            # out of loop
```

Note that in "while" loops, the programmer has more responsibility for the correct behavior of the loop. Incorrectly initializing the running index, forgetting to increase it every time, or a wrong loop termination condition will result in an incorrect computation.

"For" loops are more limited than "while" loops: with "for" loops, we can iterate over a given collection of elements, but sometimes we need our program to keep on doing something as long as some condition holds, in which case "while" loops are the only option. For example, suppose we seek all positions in a given protein where a specific restriction site of an enzyme occurs. Here is a function to do the job:

```
1 def cut_positions(seq, pattern):
2     ''' find all positions of pattern within seq '''
3     cut_sites = []
4     site = seq.find(pattern) # get index of pattern in seq, -1 if none
5     while site != -1:
6         cut_sites.append(site)
7         site = seq.find(pattern, site+1) # get next match
8     return all_sites
```

> ### Code explained
>
> The method `find` of class `str` can receive one or two parameters. In the former case, as in line 4, it returns the first index in which the input pattern string occurs within `seq`. When given two parameters, as in line 7, the second one specifies at which index of `seq` to start the search. So `seq.find(pattern, site+1)` finds the first occurrence of pattern in `seq` ignoring the first `site` positions, where `site` is the position of the previous match. The loop's condition in line 5 is `site != -1`, because the method returns −1 when no match was found.

1.8.3 Looping with `range`

Python provides an easy way to define a **range of integers** with a constant leap (this may remind you of how slicing works). The command `range(a,b)` generates all the integers between `a` and `b`, not including `b`. Adding a third parameter `range(a,b,c)` provides those integers with a "leap" of `c` positions.

```
>>> for i in range(3,7):
        print(i*2)

6
8
10
12

>>> dna = "ATTG"
>>> for i in range(0,len(dna)):
        print(dna[i]*2)

AA
TT
TT
GG

>>> list(range(3,10,2))
[3,5,7,9]
```

Furthermore, if we provide `range` with only a single parameter, the default starting index is 0:

```
>>> list(range(7))
[0,1,2,3,4,5,6]
```

Note that we can always use a simple "while" loop instead of using `range`, but often the latter is more convenient.

Here is another example, using `range`, of a function that finds the first stop codon in a given DNA sequence, taking into consideration the frame shift:

```
1 def find_stop(dna):
2     stops = ["TAG", "TAA", "TGA"]
3     n = len(dna)
4     for i in range(0, n, 3):
5         if dna[i:i+3] in stops:
6             return i
```

Code explained

The loop in line 4 jumps three positions in each iteration, starting from 0, so the values of i are 0, 3, 6, 9... (the last value depends on the length of the string). For every such value of i, the function uses slicing to extract the three nucleotides at positions i, i + 1, i + 2 and to check if this codon is a stop codon.

Exercise 17

What will find_stop return if no stop codon is found in the DNA input?

1.8.4 Iterating over Sequences – Summary

Let us summarize the three ways we have in Python to iterate over a given sequence, such as a string or a list. The first way we saw accesses the elements directly using a for loop:

```
for element in seq:
    do something with element
```

This is the most convenient way, but it is appropriate only when we wish to access merely the elements in the sequence, without any reference to their positions (indices) within the sequence. The second way uses a while loop, and iterates over the indices, through which the elements are accessed:

```
i = 0
while i < len(seq):
    do something with seq[i]
    i = i+1
```

The third way uses range to iterate over the indices as well:

```
for i in range(len(seq)):
    do something with seq[i]
```

> ### Exercise 18
>
> Write a function named `reverse_comp(dna)` that receives a DNA sequence, and returns its reverse complement,
>
> (a) By using Python's `range` to iterate over `dna` in a reverse order.
> (b) By reversing the string itself using slicing

1.9 Interactive User Input

It is common for a program to depend on some input from the user in an interactive manner, namely input that is provided by the user at runtime. As an example, recall the function `state` from an earlier section. We could ask the user to provide the room temperature at a given moment during the program's execution. The way to do this in Python is to use the built-in function `input`:

```
>>> room_temp = input("Please provide the current room temperature: ")
```

This command will make your program stand-by, with the message string displayed. The program's execution is paused until the user provides some input and presses <Enter>. Once this happens, the function `input` captures this value and returns it.

```
>>> room_temp = input("Please provide the current temperature: ")
Please provide the current temperature: 29
>>> print(room_temp)
'29'
```

29 is typed by the user at runtime

Note that the function input always returns a string. (Clearly, it has no way of knowing what type of data the user intended to provide – `int`, `str`, etc. It treats any characters typed as a string.) But we have already seen how to convert strings to integers. Now we can get the room temperature from the user at runtime, and use the function `state` as before:

```
1  def get_room_temp():
2      room_temp = input("Please provide the current temperature: ")
3      return int(room_temp)
4
5  def state(Tm, Tv, T):
6      '''return the state of material with Tm and Tv at temperature T '''
7      if T < Tm:
8          return "solid"
9      elif T < Tv:
10         return "liquid"
11     else:
12         return "gas"
```

```
>>> T = get_room_temp()
Please provide the current temperature: 29
>>> water_state = state(0,100,T)
>>> print("Water at", T, "degrees is", water_state)
Water at 29 degrees is liquid
```

1.10 Containers

The term **container** in Python refers to any element that contains "inner" elements, allowing them to be accessed and iterated over. We have seen two types of containers so far: string and list. Python has additional container types, most notably *tuples*, *sets*, and *dictionaries*, which we describe in this section.

We can roughly classify containers according to two characteristics: **mutability** and **order**. We have encountered the notion of mutability earlier: a mutable object allows changing its "inner" elements whilst keeping its memory address unchanged – lists are mutable, while strings are not. Both strings and lists are ordered containers: their inner elements are ordered by positions (starting at 0). In other words, these are **sequences** of elements. The new container types presented in this section are tuples, sets, and dictionaries (Table 1.6).

Sets and **dictionaries** are unordered collections of elements. "Unordered" means that elements have no indices, thus there is no "first," "second," or "last" elements in them. This is in contrast to lists and strings, in which elements are positioned as a contiguous sequence, and can be accessed using indices 0, 1, 2, etc. Sets in Python resemble a lot of sets in mathematics: they represent an unordered collection of unique elements (unique means that an element cannot appear more than once in a set). In particular, they support operations such as adding and removing elements, union and intersection. Elements of sets are enclosed between {}:

```
>>> s1 = {3,4,5}
>>> 3 in s1
True
>>> s2 = set()      <= empty set
>>> s2.add(3)
>>> s2.add(2)
>>> s2
{2, 3}
>>> s2.add(2)
>>> s2
{2, 3}              <= no repetitions in sets
>>> s2.discard(2)
>>> s2
{3}
>>> s2.discard(5)
>>> s2
{3}
>>> s2.union({3,4,5})
{3, 4, 5}
>>> s2.intersection({3,7})
{3}
```

Table 1.6. *Container types*

	Ordered (sequences)	Unordered
Mutable	list [1,2,"hello"]	set {1,2,"hello"}
		dict {1:"a",2:"b",3:"c"}
Immutable	str "TGAAC"	
	tuple (1,2,"hello")	

A very convenient way to remove repetitions from a list is to convert it to a set:

```
>>> list(set([7,3,3,8,7,3]))
[8, 3, 7]
```

Note that since sets are unordered, the original order of the elements in the list is generally not preserved in this process.

Dictionaries in Python are "close relatives" of sets. They are also unordered mutable containers. However, elements in a dictionary have two components: a key and an attached value. In other words, Python's dictionaries are used to map a set of keys to values. To get the mapping of a key, we specify the key inside [ ], as in the following examples:

```
>>> d1 = dict()   <= empty dictionary
>>> d1["Lys"] = "K" <= insert a new mapping, "Lys" mapped to "K"
>>> d1
{'Lys': 'K'}
>>> d2 = {"Arg":"R", "Tyr": "Y"}<= initialize a dictionary with 2 keys
>>> d2["Arg"] <= get the value that the key "Arg" is mapped to
'R'
>>> "Arg" in d2
True
>>> d2.keys()
dict_keys(['Tyr', 'Arg'])
>>> list(d2.keys()) <= get all the keys
['Tyr', 'Arg']
>>> list(d2.values()) <= get all the values
['Y', 'R']
>>> d2.pop("Arg")
'R'
>>> d2
{'Tyr': 'Y'}
```

The following dictionary maps DNA codon to the amino acids, according to the universal genetic code:

```
1  universal = {
2      'ttt': 'F', 'tct': 'S', 'tat': 'Y', 'tgt': 'C',
3      'ttc': 'F', 'tcc': 'S', 'tac': 'Y', 'tgc': 'C',
4      'tta': 'L', 'tca': 'S', 'taa': '*', 'tga': '*',
5      'ttg': 'L', 'tcg': 'S', 'tag': '*', 'tgg': 'W',
6      'ctt': 'L', 'cct': 'P', 'cat': 'H', 'cgt': 'R',
```

```
7      'ctc': 'L', 'ccc': 'P', 'cac': 'H', 'cgc': 'R',
8      'cta': 'L', 'cca': 'P', 'caa': 'Q', 'cga': 'R',
9      'ctg': 'L', 'ccg': 'P', 'cag': 'Q', 'cgg': 'R',
10     'att': 'I', 'act': 'T', 'aat': 'N', 'agt': 'S',
11     'atc': 'I', 'acc': 'T', 'aac': 'N', 'agc': 'S',
12     'ata': 'I', 'aca': 'T', 'aaa': 'K', 'aga': 'R',
13     'atg': 'M', 'acg': 'T', 'aag': 'K', 'agg': 'R',
14     'gtt': 'V', 'gct': 'A', 'gat': 'D', 'ggt': 'G',
15     'gtc': 'V', 'gcc': 'A', 'gac': 'D', 'ggc': 'G',
16     'gta': 'V', 'gca': 'A', 'gaa': 'E', 'gga': 'G',
17     'gtg': 'V', 'gcg': 'A', 'gag': 'E', 'ggg': 'G'    }
```

As can be guessed, a star ' * ' stands for a stop codon in this dictionary. We can use this dictionary to translate a gene into a protein:

```
1   def dna_translate(gene, code=universal):
2       ''' translate a DNA sequence into protein '''
3
4       prot = ""
5       gene = gene.lower() # to match lowercase format of universal
6       for i in range(0,len(gene),3):
7           codon = gene[i:i+3]
8           if codon in code:
9               prot += code[codon] # get the corresponding amino acid
10          else:
11              prot += "?" # unexpected triplets that are not codons
12      return prot
```

```
>>> dna_translate("ATGGACGGCGGC")
'MDGG'
```

Code explained

The function `dna_translate` receives two parameters: one is the gene to be translated, and the second one is a default parameter set to the universal genetic code defined earlier. If one wishes to compute the translation according to a different genetic code, then the second parameter will be given as input to the function and overload the default value.

Line 4 creates an empty string to which the amino acids will be concatenated, one by one. Line 5 makes the translation case insensitive, by transforming the sequence into lowercase, in case it was not so. The loop in lines 6–11 iterates over the gene, in leaps of three nucleotides each time. Line 7 extracts the three nucleotides starting at position `i`. Line 8 checks if this codon appears in the dictionary of the translation (since the universal code dictionary contains all 64 possible triplets, this case will occur only if the sequence contains characters that are not a, t, c, or g). In line 9, the dictionary is accessed to extract the value to which the codon is mapped. This value is a one-letter amino acid, and is concatenated to `prot`. If the membership check failed, `"?"` is appended instead.

Tuples, defined by comma separated elements inside (), are the immutable parallel of lists: they are ordered, can store any type of elements, but do not support item assignment:

```
>>> tup = (1,2,3)
>>> tup[2] = 5
Traceback (most recent call last):
  File "<pyshell#69>", line 1, in <module>
    tup[2] = 5
TypeError: 'tuple' object does not support item assignment
```

So tuples are something like "immutable lists." What do we use them for? One of the uses of tuples is as elements in sets, and keys in dictionaries: both must be immutable elements, for reasons beyond our scope here (we just note that it has to do with efficiency, of the kind discussed in the chapter on hashing). So a list can neither be an element in a set, nor a key in a dictionary. This is where tuples come to the rescue.

```
>>> s = {1,2,[3,4,5]}
Traceback (most recent call last):
  File "<pyshell#71>", line 1, in <module>
    s = {1,2,[3,4,5]}
TypeError: unhashable type: 'list'
>>> s = {1,2,(3,4,5)}
>>> s
{(3, 4, 5), 2, 1}
```

Pay attention that the order of the items in the set was changed in the last running and not by mistake. It is because a set is an unordered collection. Python has an inner order for it, but for us, as outside users, the order seems random.

1.11 Files

Files are used to store long-term data. Unlike variables defined during a program's execution, files can store much larger amounts of data, and these data do not get lost when the program terminates, not even when the computer shuts down. When biologists want to store the genome of an organism or the expression data of various genes in healthy and mutant cells, they store the data in files. The data in the files are later being read by programs to perform various analyses.

Working with text files in Python is very easy. Other formats (such as Excel spreadsheets or xml files) can also be handled, but involve more details and we do not deal with them here. The fundamental text file operations are **opening** and **closing** a file, **reading** its content, and **writing** data to it. Here is an example, in which the content of one file is copied to another, except for the first line:

```
>>> my_file = open("./yeast_genome.txt")  <= ./ for current folder
                                           <= ../ for parent folder
>>> out_file = open("../out.txt", 'w')     <= 'w' for writing, file will
                                           <= be created if does not exist

>>> first_line = my_file.readline()        <= read until '\n' encountered
>>> genome =  my_file.read()               <= read rest of file

>>> print(genome, file=out_file)           <= write to file specified

>>> my_file.close()
>>> out_file.close()
```

> **Code explained**
>
> Python function `open` is used to open files. The name of the file is given as a string parameter. If the file is located at the same directory as the Python script file, the file's name is enough. Otherwise, you can provide the full path to the location of the file. "./" means the current directory, and "../" points to the parent directory. To write data to a file, it first has to be opened with a special flag 'w'. Then, a simple `print` command, directed to the file, will do the job.
>
> The function `read` is used to load the content of the file. Assuming the file is a simple text file, the content is loaded as a single string type value, and can be stored in a variable. When the file contains line escape characters (the special "end of line" character '\n'), we may prefer to read its content line by line and handle each separately, using the `readline` function.
>
> Closing a file terminates the communication with the file system. It is a good habit to close a file once done with it. This is especially important when writing data to the file: until the file is not closed, it is not guaranteed that all the data were actually written to it. This is because writing to a file is an expensive operation, and the computer occasionally prefers to delay it, buffer the data, and write them to the file only when enough data are accumulated.

1.12 Libraries

We conclude this chapter by presenting the use of **libraries**. Many useful objects and functions in Python come as part of libraries, also termed **packages**. A library can be either a standard part of Python or written by third party developers. For example, Python has a `math` library, containing various mathematical functions; a library for random sampling and probability called `random`; a library for time measurements called `time`; and many more. To use a library, we first need to "import" it into the current running environment. Then, we can access all that it has to offer, as shown in the following examples:

```
>>> import random
>>> random.choice("ATCG")        <= function choice of package random
'A'                              <= picked an element uniformly at random
>>> random.choice("ATCG")        <= pick again
'G'
>>> random.choice(range(10))     <= randomly pick from the range 0-9
8

>>> seq = ""
>>> for i in range(10):
        seq += random.choice("ATCG")  <= random DNA sequence of length 10

>>> seq
'AGCCACCCGA'

>>> random.random()              <= random decimal number from [0,1)
0.7601425767073609
>>> if random.random() <= 0.3:  # if the number picked is between 0-0.3
        print("30% chance this will be printed")

>>> lst = [1,2,3,4]
>>> random.shuffle(lst)          <= randomly permute the list
>>> lst
[2, 1, 3, 4]
```

Exercise 19

Complete the following function `deletion_mutation(dna)` that receives a DNA sequence, and randomly deletes one of its bases.

```
import random
def deletion_mutation(dna):
    i = _____        # position to delete
    new_dna = _____
    return new_dna
```

Additional library functions will be presented later, when used. We refer the reader to Python's documentation to explore more packages and their functionality.

Reflection

Variables, operators, conditionals, loops, containers, files... we hope you are not too overwhelmed with the amount of information about Python presented in this chapter.

Although the Python basics presented will be of use in the rest of this book, this does not mean you must be a Python expert to read it. Whenever you encounter something unclear in the code we present later in the book, take your time to play a bit with the code, clarifying the unclear details. This is a good point to remind you that learning a new programming language requires a lot of practice.

Please take your time to look at the problems under the "challenge yourself" section that ends this (and every) chapter. In some problems, you will be asked to use functions that you saw in this chapter. Other problems will require modifying them or even writing new functions. It is always good to remember that Python types have various useful functions that you may consider using. You will find the solutions on our website.

Challenge Yourself

Problem 1 Using dictionaries for protein representation

The following is a dictionary in Python that maps amino acids in 3-letter format to 1-letter format:

```
prot321_dict = {"GLY" : "G", "ALA" : "A", "LEU" : "L", "ILE" : "I",
                "ARG" : "R", "LYS" : "K", "MET" : "M", "CYS" : "C",
                "TYR" : "Y", "THR" : "T", "PRO" : "P", "SER" : "S",
                "TRP" : "W", "ASP" : "D", "GLU" : "E", "ASN" : "N",
                "GLN" : "Q", "PHE" : "F", "HIS" : "H", "XXX" : "X"}
```

(a) This dictionary has two problems:
1. The value for Valine ("VAL") is missing. Add it to the dictionary.
2. It has an incorrect key – "XXX." Delete it from the dictionary.
(b) Write a function `prot321(prot)` that takes a protein sequence `prot` in 3-letter representation (separated by spaces). For example: "GLN ALA GLN ILE." The function returns the protein in 1-letter representation (and no spaces). For example: 'QAQI'.

```
>>> prot321("GLN ALA GLN ILE")
'QAQI'
```

Hint: You may want to use the function `split` from class str, to separate the input protein sequence into the triplets. For example:

```
>>> "GLN ALA GLN ILE".split()
['GLN', 'ALA', 'GLN', 'ILE']
```

Problem 2 File and string basics

In the website of this book, you will find a file IME1.txt, containing the DNA sequence of the gene IME1 (of S.Cerevisiae, sequence downloaded from www.yeastgenome.org). We want you to read this sequence from the file, translate it into a protein, and compute the percentage of basic amino acids in it.

Stage A – Reading a file

The sequence is given in FASTA format. In this format, before the gene itself, there is a title row containing basic information on the gene. This row usually starts with the ">" character.

(a) Read the DNA sequence, omitting the title row, into a variable `ime1_dna`.
(b) The length of the gene should be 1083 nucleotides. Check if this is the case (hint: no!).
(c) Check if the sequence contains only the nucleotides `A`, `T`, `C`, `G`. If not, what other characters are in that string?
(d) Another way to answer the previous section is by using a structure called **list comprehension.** This is just another convenient way to form lists in Python.
The general syntax is: [*expression* for *variable_name* in *sequence* if *condition*]
Now, run this command:

```
>>> [char for char in ime1_dna if char not in "ATCG"]
```

(e) Clean the sequence from non-ATCG characters. Use the method `replace` of class str.
(f) Check again the length of the gene. It should now be 1083.
(g) Verify that the length of the gene is divided by 3, using the modulo (%) operator.
(h) What are the last three nucleotides in the sequence? Use string slicing.
(i) Verify, using the dictionary 'universal' (that maps codons to amino acids), which we saw earlier, that the last codon is a stop codon.

Stage B – Translation and analysis

(a) Use the function `dna_translate` to translate the IME1 gene into a protein. Verify that the length of the protein is what you expect.
(b) Note that the last character in the protein is '*', which represents a stop codon. Remove this character from the string (you can use slicing or the method `replace`).
(c) What is the percentage of basic amino acids (lysine – K, histidine – H, arginine – R) in the protein?

Problem 3 Estimating silent mutation probability

Let us define a *complete random mutation*: given a DNA sequence, a randomly picked nucleotide in it will change to one of the other three possibilities, also picked at random. For example, given "ATTCGG," assume the first position was picked for the mutation, then A will be changed to either C, G, or T at random with equal probabilities.

In this problem, we will empirically estimate the probability for a complete random mutation being silent (that is, not affecting the protein sequence it encodes).

Following is a function `rand_mut`, which takes a DNA sequence, and returns this sequence after a complete random mutation has been applied to it.

```
1 def rand_mut(dna):
2     ''' return a new dna with a single random mutation '''
3     dna = dna.lower()
4     position = random.randrange(0, len(dna))
5     nuc_targets = list("atcg")
6     nuc_targets.remove(dna[position])
7     target = random.choice(nuc_targets)
8     new_dna = dna[:position] + target + dna[position+1:]
9     return new_dna
```

For example:

```
>>> rand_mut("ATTCGG")
'GTTCGG'
>>> rand_mut("ATTCGG")
'ATTCAG'
```

(a) Complete the following function `is_silent`, which applies a random mutation in its input DNA sequence, and returns `True` if this mutation was silent, otherwise `False`.

```
1 def is_silent(dna):
2     new = rand_mut(dna)
3     return_____
```

(b) Complete the function `prob_silent`, which takes a DNA sequence, and the desired number of simulations required, and returns the fraction of simulations that yielded a silent random mutation. Note that each simulation generates a single random mutation at the original DNA, that is, the mutations are not accumulated.

```
1 def prob_silent(dna, num_simulations):
```

```
>>> prob_silent("atc", 10000)
0.2157
>>> prob_silent("atc", 10000)
0.2238
```

Note the different results in the last two executions. This is expected as each time different mutations are randomly generated.

(c) The probability for a silent mutation in the codon 'atc' is 2/9 (why?). Verify this experimentally by enlarging the number of simulations, which increases the significance of the output.

(d) Let us define a random DNA as a DNA sequence in which each position is randomly and uniformly assigned one of the four nucleotides. Write a function `rand_dna(n)` that generates a random DNA of length n.

(e) Estimate the silent mutation probability in a random DNA of length 1000.

(f) Estimate the silent mutation probability in the DNA sequence of the S.Cerevisiae IME1 gene from the previous question.

(g) Discussion: Can you explain the results? Are they surprising?

2 Efficiency Matters – Gentle Introduction to Complexity

In this chapter, you will encounter one of the most fundamental notions in CS – complexity of computations. We take a rather non-formal approach in presenting this topic, with an emphasis on intuition. Complexity of algorithms has two flavors – time complexity, which is related to running time, and memory, or space complexity, which reflects the memory allocation requirements of an algorithm. In this chapter, and in most other chapters as well, we will focus primarily on time complexity (except for some cases, in which memory poses a real limitation). You are not expected to become an expert in analyzing the complexity of algorithms. However, you should be able, by the end of this chapter, to understand how running time depends on an algorithm's input size and to estimate empirically an algorithm's actual running time. Later in the book, in chapter 11 (Mission Infeasible), we will get acquainted with the notion of complexity from another perspective – computational problems that can be solved in principle, but for which no efficient solutions are known.

2.1 Running Time of Algorithms

Suppose you wrote a program in Python in order to solve some computational problem. How would you evaluate the quality of your solution? Quality is a rather abstract notion, and indeed can be measured in various ways. You may consider using the number of lines (or characters) in your code, under the assumption that "shorter is better". However, this is actually a bad idea. Shorter code does not necessarily imply a better algorithm. You could test the correctness of your code on various inputs, to see if it contains any tricky bugs. You may ask your friends if they understand what your code does, to get some impression of its clarity and comprehensibility ... but possibly the most important criteria are the efficiency of your algorithm. In computer science, efficiency relates mainly to two resources – **running time** and computer **memory**.

To measure how efficient an algorithm is, we can simply measure its running time, in seconds, minutes, hours, etc. Python's library `time` contains a function named `perf_counter()`. This function returns the current "processor time", expressed as a floating point number, in seconds. We can use it to measure the running time of any piece of code, as follows:

```
1 import time
2 def stopper():
3     t0 = time.perf_counter()
4     i = 0
5     while i<10**6:
6         i = i+1
7     t1 = time.perf_counter()
8     return t1-t0
```

 Code explained

In line 3, we set the "stopper" by calling `time.perf_counter()`. The processor time at this instance is recorded as the variable `t0`. In line 7, a second time measurement is carried out, and the result is recorded as the variable `t1`. Between these two lines we inserted the code whose execution time we want to measure. Specifically, here we measure the time it takes for a simple loop with 10**6 iterations to run.

```
>>> stopper()
0.1417096476839597
>>> stopper()
0.1404819005998592
>>> stopper()
0.1433337478460146
```

Note how three calls to the same program return slightly different values. In addition to inherent limitation of precision in measuring time by the computer, often the same piece of code may take different amounts of time to run, due to the technical details of the underlying execution (the processor, the operating system, and various other parameters may have influence). But for our rather basic needs, the precision given by `stopper` is more than enough.

Exercise 1

Change the loop to execute 10**7 iterations, and run it three times. What is the running time now? How does it compare to the 10**6 iterations?

Exercise 1 shows us something that was to be expected: the running time depends on the number of iterations. More generally, given some input, a function's running time depends on the input, and specifically on its size. Larger inputs normally require more time. For example, computing the GC content of a genome will take longer for larger genomes, since the algorithm will have to examine more bases.

We could measure the running time of algorithms on specific inputs. This, however, has two main disadvantages. First, actual running time on a specific input is likely to change between computers (your new laptop is probably faster than an old PC in your town's public library, and slower than NASA's computers.). Even on the same computer, running time may change significantly. (Try opening your browser at the same time when an anti-virus software is performing its weekly scan. Isn't that annoyingly slow?) Second, measuring running time on a specific input may not tell us very much about the behavior of the algorithm on other inputs.

Instead of measuring actual running time on specific inputs, computer scientists came up with a more general and scientific way to measure the efficiency of algorithms in terms of running time, called **time complexity**. First, instead of measuring time, they count the number of operations performed by the algorithm, a number which is platform invariant. The number of operations reflects an algorithm's inherent efficiency, rather than the platform's speed. In addition, instead of choosing specific inputs, the number of operations is described and analyzed as a function of the input size. In the next sections, we explore this fundamental idea in more detail.

2.2 Linear and Quadratic Time Complexity

Consider the following piece of code:

```
1 def linear(L):
2     for i in range(len(L)):
3         print(L[i])
```

Given a list, L, as input, this function simply iterates over the input list and prints each element. For a list of length n, there are exactly n printouts. Suppose you executed this function on a list of length 1000, and it took about 1 second. How long should we expect it to take for a list twice the size (2000)? Probably about 2 seconds (if we ignore the sizes of the printed numbers). In other words, the running time increases linearly with the input size. This is true, regardless of the actual time it takes for each list size. Graphically, plotting running time as a function of the input size will appear as a straight line. We say that the running time of this function has a **linear rate of growth** (or grows linearly) (see Figure 2.1).

What would happen if we printed only every other element in the list?

```
1 def linear_too(L):
2     for i in range(0,len(L),2):
3         print(L[i])
```

Here, running time should reduce to roughly half. But if we are interested in how it grows when n grows, the answer remains the same – linear. To see why, note that about $n/2$

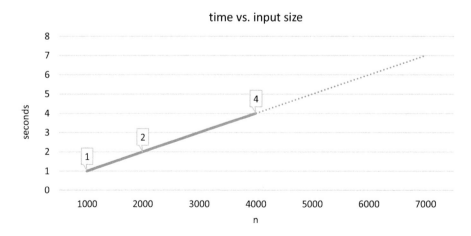

Figure 2.1. Time versus input size for linear

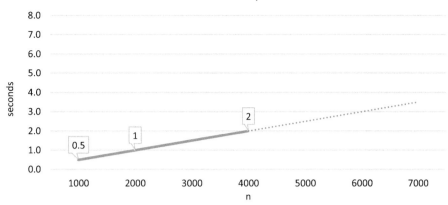

Figure 2.2. Time versus input size for `linear_too`

elements are printed. When we increase n from e.g., 1000 to 2000, the number of printouts increases from about 500 to 1000 (see Figure 2.2).

Usually, when we examine an algorithm's time complexity, we will care about the rate of growth of the running time as a function of the input size. Of course, actual running time is also very meaningful. But actual running time alone is not enough, and it depends on various parameters such as the speed of the computer hardware, the programming language used, the exact details of the implementation, and other external factors, such as the load on the computer. Therefore, to measure an algorithm's efficiency (rather than a computer's speed), we need to inspect how running time is expected to increase with the input size. In the previous example, both n and $n/2$ depend linearly on n. So do $1000n$ and $\frac{17}{2}n + 32$. When we plot these expressions as a function of n, all yield a straight line, regardless of the exact constants in the expression.

It is important to emphasize: n is indeed twice as large as $n/2$, and 1000 times smaller than $1000n$. So we do not claim that algorithms that perform this amount of operations (e.g., printouts) have the same running times. We only say that they have the same complexity, which refers to the rate of growth of their running time. The rate is not affected by these constants. We denote the running time of functions with linear time complexity by $O(n)$, read as "Oh" of n, where n is the input length.

Exercise 2

A palindrome is a string that is read the same forward and backward. For example, the strings "ABBA," "kayak," and "racecar" are English palindromes. Write a function `is_pal(st)` that takes a string and returns `True` if it is a palindrome, and `False` if it is not. Your function should work in linear time complexity, where n is the number of characters in `st`. Specifically, try to limit yourself to only about $n/2$ loop iterations.

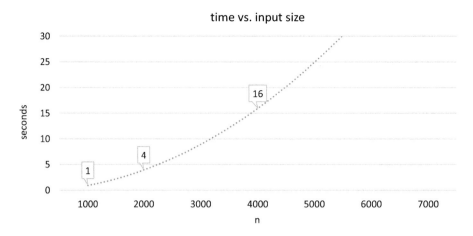

Figure 2.3. Time versus input size for `quadratic`

Now let us compare the above to the following function:

```
1 def quadratic(L):
2     for i in range(len(L)):
3         for j in range(len(L)):
4             print(L[i], L[j])
```

This time, the function prints every pair of elements in the list. For a list of length n, there are n^2 such pairs. Suppose, again, that it took this function 1 second to run on a list of size 1000. How much longer do you expect it to run on a list of size 2000? In other words, how does running time increase when n increases? When n is doubled, n^2 increases four times. So, we expect running time for a 2000 long list to be roughly four seconds. This rate of growth is termed **quadratic** (see Figure 2.3).

What would change if we let the index j start at i rather than at 0?

```
1 def quadratic_too(L):
2     for i in range(len(L)):
3         for j in range(i,len(L)): # note the change
4             print(L[i], L[j])
```

Of course, the running time of this function is expected to be smaller than the previous one. It will run in approximately half the time, because about half the pairs (i, j) are ignored (only pairs where $i \leq j$ will be printed. See Table 2.1).

But note that when increasing n (the size of the list), the running time is still expected to increase quadratically. To see why this is true, we can count the number of pairs printed. For $i = 0$, we have n printouts: (0,0), (0,1), ..., (0, $n-1$). For $i = 1$, we only have $n - 1$ printouts: (1,1), (1,2), ..., (1, $n-1$). In total, we have $n + (n-1) + (n-2) + \ldots + 1 + 0$ printouts (the last term, 0, refers to the case where $i = n - 1$, so we do not even enter the inner loop). This is an arithmetic series, and we have $n + (n-1) + (n-2) + \ldots + 1 + 0 = \frac{n(n+1)}{2} = \frac{n^2}{2} + \frac{n}{2}$. The "dominant" term in this expression is quadratic, and indeed this expression increases roughly by a factor of 4 when n is doubled. Other examples for quadratic expressions

Table 2.1. *This loop iterates only over the pairs i, j that appear in the colored cells. The orange diagonal includes pairs where i = j, and the pink triangle above it is where i < j*

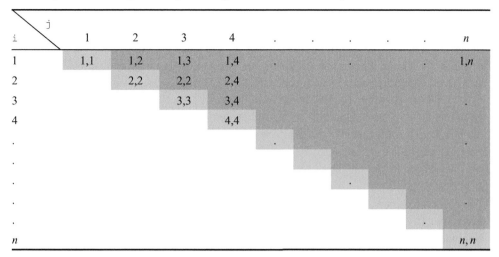

i \ j	1	2	3	4	.	.	.	.	.	n
1	1,1	1,2	1,3	1,4	.		.			1,n
2		2,2	2,2	2,4						
3			3,3	3,4						.
4				4,4						
.					.					.
.						.				
.							.			.
.								.		
.									.	
n										n, n

are $1000n^2$, $0.001n^2 + 300n$ and $2.5n^2 - 25,000$. If an algorithm performs a number of operations that is equal to such expressions, we expect its running time to increase roughly four times when n, the input size, is doubled. This behavior occurs asymptotically, regardless of the exact constants and lower order terms (such as $300n$ or $-25,000$).

Suppose we have two alternative algorithms for the same problem, one performs n operations, while the other n^2. Obviously, we prefer the former, since $n < n^2$ for $n > 1$. That was easy. But what about $1000n$ compared to $0.001n^2$? For values of n up to 10^6, $1000n$ is actually larger than $0.001n^2$. For larger values, however, $1000n$ "wins." This can be generalized: for every two algorithms, one that performs a linear number of operations as a function of its input size, and the other with a quadratic number of operations, for a large enough input size, the linear algorithm is more efficient.

This does not mean that we will always prefer the linear algorithm over the quadratic one. For small inputs, the constants and lower rate terms may need to be carefully taken into consideration. But for large enough inputs, the linear algorithm will indeed be superior. The exact input size from which this happens depends, again, on the constants and lower rate terms. A different consideration is which algorithm may be easier to code, and consequently the implementation less prone to errors.

2.3 The O Notation

Computer scientists have a compact way to denote an algorithm's complexity. They use the so-called **O notation**. For example, linear complexity is denoted $O(n)$ (read, "Oh of n"), while quadratic complexity is $O(n^2)$ ("Oh of n square"). So we can simply say that the function linear we saw "runs in $O(n)$ time" or "has $O(n)$ time complexity." Similarly, the function quadratic "runs in $O(n^2)$ time" or "has $O(n^2)$ time complexity."

The O notation is very convenient when we want to describe and compare the efficiency of algorithms. Instead of talking about the exact number of operations, which is often rather complicated to compute and may depend on many implementation details, we can classify algorithms to **classes of complexity**. These classes, such as linear or quadratic, refer only to how running time is expected to increase asymptotically, with the input size. This measure is a meaningful way to describe the time efficiency of algorithms. The lower the complexity is, the larger the inputs we expect to be able to handle within a reasonable time. So $O(n)$ is considered more efficient than $O(n^2)$. Intuitively, the O notation helps us focus on the "dominant" part of a mathematical expression describing the number of operations performed by an algorithm. By "dominant," we mean the part that has the biggest effect on the running time expression, at least for large ns, eliminating the less dominant parts of the expression.

The O notation has its limitations, however. Sometimes, the exact number of operations is still very important. An algorithm that performs n operations on an input of size n will run 1000 times faster than one with $1000n$ operations (assuming each single operation takes the same amount of time). When analyzing the efficiency of algorithms, we may not want to ignore the "fine details" and actually count the exact number of operations (or at least a specific type of operation, such as printouts or comparisons). In such cases, the O notation is too coarse as a measure of complexity. Put differently, complexity classes tell us approximately how fast or slow an algorithm will run on larger inputs but may ignore important details about the actual running time for a given input.

A good way to estimate efficiency of algorithms is to go both ways: analyze the asymptotic time complexity in terms of O notation, to estimate whether our algorithm is expected to handle large inputs well, as well as measure actual running time to get an absolute measure of efficiency. In this book, we will choose one or both of these approaches, depending on the need.

2.4 Best-Case, Worst-Case

The examples we have seen so far all exhibited the following property: their complexity depended merely on the size of the input. In some cases, complexity may depend also on the input itself. For example, if the input is a list, then the content of the list, in addition to its length, can affect the time complexity. A simple example for that is a search algorithm. Suppose we have a list L and an element x, and we want to find if x is in L. We may employ a simple loop to solve this search problem:

```
1 def search(L, x):
2     for item in L:
3         if item == x:
4             return True
5     return False
```

Given a list L of size n, how many comparisons (line 3) are executed by this algorithm? If x is a member in L, the answer depends not only on n, but also on the position in L where x is located. So when analyzing time complexity, we should consider the position as well. To that end, it is convenient to separate the analysis into the **best-case** and the **worst-case** scenarios. In the best case, when $L[0] == x$, only one comparison is needed. In the worst case, when x is not a member of L, we need n comparisons. So the time complexity is $O(1)$[1] in the best-case and $O(n)$ in the

---

[1] $O(1)$ means constant complexity, which does not increase with input size at all.

worst-case scenario. Both these extremes may sometimes be very rare, and are far from a "representative" or a common input. In such cases, computer scientists often consider the expected, or average time complexity. We remark that the average case analysis is usually harder to analyze than the worst (or best) case. Worst-case analysis is the one considered in most cases.

2.5 Binary Search and Logarithmic Time Complexity

There are many other complexity classes in addition to linear and quadratic ($O(n)$ and $O(n^2)$, respectively). One important such class is the **logarithmic** time complexity class, $O(logn)$. A notable example for this class is the binary search algorithm. Let us first describe it and then analyze its time complexity.

Binary search is a widely applicable technique, best known for its application for searching an element in a *sorted* list. Suppose we have a sorted list, L, and an element, x, and we want to check if x is in L or not. If the list is sorted, one can do much better than merely checking each item in L, sequentially, comparing it to the desired x. Binary search starts at the middle element of L, and compares the item at that location to x. This comparison may have three possible outcomes. If the two are equal, the search can terminate, because we have just found x in the list; if x is smaller than the middle element in L, the search can be narrowed down to the lower (left) half of L, because x cannot be found in the upper half anymore. Similarly, if x is larger, the search can be narrowed down to the upper (right) half. The following code implements this approach. The function `binary_search` assumes the list L is sorted.

```
1   def binary_search(L, x):
2       left, right = 0, len(L)-1
3       while left <= right:
4           mid = (left+right)//2
5           if L[mid] == x:
6               return True
7           elif L[mid] > x:
8               right = mid-1
9           else:
10              left = mid+1
11      return False
```

Code explained

Line 2 makes a "simultaneous assignment." The variable `left` is assigned 0, and `right` is assigned `len(L)-1`. The condition in the third line means "the interval of the list `L` in which the search is restricted to is non-empty." When `left == right`, this interval is of size 1. Line 4 computes the middle index of that interval, each time we enter the loop. If `len(L)` is even, the middle index is not an integer, in which case `mid` is rounded down (e.g., a list of length 10, then $mid = (0+9)//2 = 4$). The function terminates and returns `True` the first time it encounters an element in the list equal to `x`. Lines 7–10 narrow down the search interval as explained above. If the function reached line 11, this means no element equal to `x` was found and the search interval became empty, thus `False` is returned.

> **Exercise 3**
>
> (a) Track the execution of `binary_search` on the list L = [1,3,4,5,9,12,13, 26,29,32], and the following values of x: $x = 3$, $x = 6$ and $x = 33$. For the tracking, add a print command at the beginning of each iteration to observe the value of the variable `mid`.
> (b) What happens when the input list is *not* sorted? Can the function mistakenly return `True` when x is not in L ("false positive")? Can it mistakenly return `False` when x is in L ("false negative")?

Binary search executes way fewer than n iterations even in the worst case. In each iteration, the current range of possible indices is halved. So after 1 iteration we narrow the range down to about $n/2$ elements; after two iterations – to about $n/4$, and so on. How many iterations are needed in the worst case? The answer is the number of times one can divide n by 2 (rounded down each time) until getting a result below 1. This is essentially the logarithm to the base 2 of n, denoted $log_2 n$. In terms of O notation, such an algorithm has time complexity O($\log n$). Logarithmic time algorithms are much more efficient than linear, let alone quadratic ones. Note, however, that the list has to be sorted before running a binary search, and usually sorting is 'not for free'.

> **Exercise 4**
>
> (a) What is the number of iterations of `binary_search` on a list of size $n = 1000$ in the worst case?
> (b) How many more iterations will be needed for $n = 2000$?

2.6 Exponential Complexity and Tractability

Algorithms that run in logarithmic, linear, quadratic, or any higher polynomial (e.g., n^3, n^4) time complexity, are termed **tractable**. This term refers to the fact that such algorithms are expected to terminate and return an answer within a reasonable time, even on fairly large inputs. Of course, at some point, the input size may be too large and running time too long. But when the complexity is lower, it will take a larger input to cause such behavior. Table 2.2

Table 2.2. Tractable and intractable complexity classes

Tractability	Complexity class	Maximal n for which running time <= 1 sec
Tractable	$log_2 n$	$n = 2^{(10^{10})} = 2^{10,000,000,000}$
	n	$n = 10^{10} = 10,000,000,000$
	n^2	$n = \sqrt{10^{10}} = 10^5 = 100,000$
Intractable	2^n	$n = log_2(10^{10}) \approx 33$

shows the maximal input size that could still be handled by a modern computer within 1 second, for algorithms of various complexity classes. We assume our computer can perform 10^{10} (ten billion) operations per second.

As you can see, tractable algorithms handle relatively large inputs well. The complexity class $O(2^n)$ is a different story, though. It is termed **exponential** time complexity, and belongs to the **intractable** complexity classes (together with, e.g., 3^n, $n!$, and many more). When we have an exponential time algorithm, we can normally run it only on very small inputs. Running time simply increases too fast with the input size. This is not merely a theoretic notion. Intractable algorithms are typically impractical for input sizes beyond a few dozen, as can be seen in the table above.

As a simple example for an exponential time algorithm, consider the problem of printing all genomic sequences of length n for a given positive integer n. Since there are 4^n such sequences, any algorithm, no matter how efficiently it moves from one sequence to the next, will print $n \cdot 4^n$ characters in total. So there is no way to avoid the exponential time complexity. For values of n over about 30, such an algorithm is impractical.

Exercise 5

Suppose your computer can write 10^{10} characters per second to a file. How much time will it take for an algorithm to print *all* genomic sequences of length $n = 30$? Ignore the time it takes for that algorithm to compute those sequences (move from one to the next).

2.7 Useful Tips

We conclude this chapter with some practical advice that may assist you when evaluating your algorithm's efficiency.

Most built-in functions do not 'work for free': note that when you use Python's built-in functions, they usually do not perform a constant number of operations, but rather a number that depends on the input size. For example, computing the minimum of a list of numbers, using Python's `min` function requires checking every element in the list. This, for a list of length n, takes n operations, implying $O(n)$, or linear time. The same goes for summing up a list, reversing a list, etc.

Sorting: Python has a built-in function for sorting, called `sorted`. It generates a sorted copy of the input sequence. There are many sorting algorithms, based on various approaches. Without getting into this widely studied topic, we just mention that sorting a list of length n normally takes $O(n \log n)$ time. This is only slightly worse than linear time complexity (for very large values of n, you may notice a difference).

Slicing: one of the common operations used on biological sequences (as well as lists and strings in general) is slicing, e.g., `genome[start:end]`. It is worth remembering that this operation actually generates a copy of the sliced substring. Thus, for a substring or sublist of length m, slicing involves copying m characters, and requires $O(m)$ time complexity.

> ### Exercise 6
>
> What is the time complexity of the following code, as a function of *n* (where `st` is a string)?
>
> ```python
> def func(st):
>     n = len(st)
>     for i in range(n+1):
>         print(st[0:i])
> ```

Multiple inputs: algorithms often have more than a single input. In such cases, the complexity may depend on several input sizes. A simple example:

```python
1 def concat(st1, st2):
2     st3 = st1+st2
3     return st3
```

This function takes two strings as input, and generates a new string `st3`, which is the concatenation of its inputs. If the length of `st1` is n_1 and that of `st2` is n_2, then `st3` is $n_1 + n_2$ characters long. Since the function needs to copy $n_1 + n_2$ characters to create `st3`, its time complexity is $O(n_1 + n_2)$. What does this mean? It means that the time complexity is linear in the sum of lengths of the inputs. When this sum is twice as large, the running time is expected to double.

Dealing with extremely large inputs: in the biological context, we often work with, e.g., genomes or other sequences that may be millions or billions of characters long. In such cases, even quadratic time algorithms may be too inefficient. Thus, we should aim at devising faster algorithms, e.g., linear time, $O(n)$, or slightly slower, e.g., $O(n \log n)$. This is not always possible, however, as some problems simply require larger complexity to solve.

Trust but check: time complexity analyses may sometimes seem to contradict actual running time measurements for small size inputs. For example, it could happen that an $O(n)$ time algorithm will run slower than an $O(n^2)$ one for small values of *n*. The source for such a scenario lies within the constants hidden in the O notation. Suppose that your linear time algorithm performs $10^6 n$ operations, and the quadratic one just n^2 operations (we assume a single operation takes the same amount of time in both algorithms). Then, up to $n = 10^6$, the linear time algorithm requires more operations. So when comparing the efficiency of algorithms, the characteristic input sizes should be also considered, when we know them.

Reflection

In this chapter, we focused on the fundamental notion of complexity of algorithms. Even without a formal complexity analysis skill, it is a good idea to have some sense of how an algorithm's complexity depends on the input size, and whether the best and worst case differ significantly in this respect. When an algorithm runs too slowly, actual running time measurements can assist in estimating the running time. Then, we can decide if being a bit more patient is justified, or whether we should try looking for a better solution.

In this chapter, we focused on time complexity. Memory consumption requirement poses another major limitation for the execution of algorithms. More memory may often be used to save time, as we will see in the next chapter. When an algorithm consumes too much memory, your computer may not be able to execute it at all, thus the time saving is irrelevant. So both time and memory should be taken into consideration when devising a solution to a computational problem. Often (but not always) there is a tradeoff between these two resources.

Efficiency should always be taken into consideration when planning an algorithm. Sometimes an inefficient algorithm may be practically useless, if it requires for example a week to run, or fails due to a memory overflow.

Challenge Yourself

Problem 1 The repeating *k*-mer problem

Suppose we are given a string *st* and an integer *k*, and we ask whether *st* contains a substring of length *k* (*k*-mer) that repeats itself (appears at least twice).

For example, for:

st = "CCTAGTCCA" and
k = 2

there is a repeating 2-mer "CC". But there is no repeating 3-mer (check this out!).

Suppose we allow overlaps between substrings. For example, *st* = "AGAGA" contains a substring of length 3 that appears twice – "AGA," and *st* = "TTTTT" has "TTTT" twice (and "TTT" three times).

This exercise investigates several solutions to this problem. We start with the naïve solution. While such a solution may not be too efficient, it is easy to implement, understand, and modify. When the input size is not too large, this naïve solution may even terminate within a reasonable amount of time. The following function solves this problem naively:

```
1 def repeating_substring_naivel(st, k):
2     n = len(st)
3     for i in range(n-k+1):
4         for j in range(i+1, n-k+1):
5             if st[i:i+k] == st[j:j+k] :
6                 return st[i:i+k]
7     return None
```

(a) Explain why `j` starts at `i+1` and not at 0.
(b) Suppose we wanted to avoid overlaps. Where should `j` start then?
(c) How many iterations in total does this algorithm perform in the worst case, as a function of *n* and *k*, where *n* is the length of *st*? Assume *k* is much smaller than *n*. What about the best case?

(d) Run the function on the *Mycobacterium tuberculosis* genome (you can find it on our website), to find repeating substrings of length 70. Does your execution terminate within 2 minutes?

(e) Estimate the total running time expected for the execution in the previous section to terminate, by adding "diagnostic printouts" to the function.

(f) The following alternative solution makes use of Python's 'in' operator. Explain why this solution is incorrect.

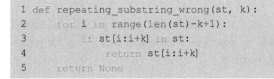

```
1 def repeating_substring_wrong(st, k):
2     for i in range(len(st)-k+1):
3         if st[i:i+k] in st:
4             return st[i:i+k]
5     return None
```

(g) Now let us consider another correct solution, based on the method `count` of class string. Instead of the inner loop, we can simply count how many times `st[i:i+k]` appears in `st`. If it occurs more than once, we have a repeating substring. Complete the following function based on this idea:

```
1 def repeating_substring_naive2(st, k):
2     for i in range(len(st)-k+1):
3         if _____:
4             return st[i:i+k]
5     return None
```

(h) What is the time complexity of this solution?

(i) Run the code on the *Mycobacterium tuberculosis* genome, to find repeating substrings of length 70. Does the execution terminate within 2 minutes this time?

In the next chapter on hashing, we will get back to this problem and solve it more efficiently. The execution on *Mycobacterium tuberculosis* genome with $k = 70$ will terminate after a second or so!

Part II: Sequences

– "String along."

Much of the initial work in bioinformatics was centered around biological sequences of medium lengths, such as genes and protein sequences. Typical problems were finding common genomic motifs, e.g., conserved sequence motifs in the promoters' region of a gene family, or looking for sequence similarity among proteins of different species. These tasks, and many others, have significant biological consequences. Yet the key to their solution lies in the so-called area of stringology, which is the computer science jargon for the study of string algorithms and properties.

What is a string in the first place? We have some alphabet, which is a finite collection of characters. For example, the English alphabet, the Greek alphabet, the binary alphabet (2 characters), the DNA nucleotides alphabet (4 characters), or the amino acid alphabet (20 characters). A string is a finite sequence of characters from the appropriate alphabet. For example, "TATA" is a string of length 4 over the DNA alphabet. "Andwhataboutthisstring?" over the English alphabet? It is a bit hard to read, is it not? Try this version, "And what about this string?", which goes to show that space is a legitimate character, even an important one.

What operations are we usually considering within the realm of string algorithms? The simplest ones are determining the length of a string and counting the number of occurrences of a character or characters in a string. Even these simple operations can give rise to meaningful results. For example, suppose we are given a string over the DNA alphabet. Counting the number of Cs plus Gs, and dividing it by the length of the whole string, results in the GC contents. More advanced operations are reversing a string or generating the reverse complement of a string over the DNA alphabet.

Storing biological strings and retrieving biological strings is a subject of great practical importance. When dealing with long strings, like whole proteomes or whole genomes, the efficiency of these operations becomes a crucial issue. For example, if we are dealing with the human genome, which is 3 billion letters long, we cannot perform operations whose complexity is quadratic in the genome length. The number of operations will simply be way too long to complete during our life time. Important data structures that come in handy for such tasks are **hash tables**, which are discussed in chapter 3. Python's *set* and *dictionary* are implemented, behind the scenes, as hash tables, which is why they enable very efficient storage and retrieval.

Chapter 4 deals with another string-related topic: **regular expressions**. Regular expressions describe templates of characters, which define families of strings. For example, AG*TT describes all strings that start with an A, followed by any number of G (including 0), and then two T's. Regular expressions can describe more complex patterns as well. They are mostly used for searching patterns within a long text, and possibly replacing them by a different string. Many programming languages, including Python, have a built-in mechanism for expressing and utilizing regular expressions.

Before we start, a comment on computational biology and stringology. String algorithms were a major part of computer science almost from day one. The fact that the initial interface between biology and computer science dealt with biological sequences, made the interdisciplinary cooperation smoother. Later, more complex biological structures, such as protein 3D structures, interaction networks, biological images, and more, joined the show. By then, the cooperation was solid enough to withstand these more complex objects, some of which are the topic of later chapters in this book.

3 Sets, Dictionaries, and Hashing

In many contexts, we are required to handle a large collection of objects in a way that supports inserting a new object, finding if an object is present, and possibly deleting an object. These operations typically appear in a series of arbitrary length. We want all these operations to be done as efficiently as possible. Consider a search engine (and its underlying infrastructure) like Google or Bing. One makes a query (e.g., "who is the King of Asteria") and gets a response in about 945,000 results (0.44 seconds). One of the basic techniques behind such efficient implementations of search is called **hashing**.

In this chapter, we introduce and explain hashing in general, and present its built-in implementation in Python in particular (sets and dictionaries). We will use strings and substrings as our primary example of objects to demonstrate hashing. And we will employ an important problem in computational biology – the **common substring problem**.

By the end of this chapter, the reader will have a good understanding of how hash tables enable efficient solutions for some problems, and know how to use Python's sets and dictionaries efficiently for various other problems.

3.1 The Common Substring Problem

3.1.1 Description of the Problem

Suppose we have two genomes, and we want to check if they both contain an identical region of a specified length. As a concrete example, consider the following strings, which are small parts of two (much longer) bacterial genomes.

Mycobacterium tuberculosis

```
TTGACCGATGACCCCGGTTCAGGCTTCACCACAGTGTGGAACGCGGTCGTCTCCGAACTTAACGGCGACCCTAAGGTTGACGACGG
ACCCAGCAGTGATGCTAATCTCAGCGCTCCGCTGACCCCTCAGCAAAGGGCTTGGCTCAATCTCGTCCAGCCATTGACCATCGTCG
AGGGGTTTGCTCTGTTATCCGTGCCGAGCAGCTTTGTC...
```

Salmonella enterica

```
AGAGATTACGTCTGGTTGCAAGAGATCATAACAGGGGAAATTGATTGAAAATAAATATATCGCCAGCAGCACATGAACAAGTTTCG
GAATGTGATCAATTTAAAAATTTATTGACTTAGGCGGGCAGATACTTTAACCAATATAGGAATACAAGACAGACAAATAAAAATGA
CAGAGTACACAACATCCATGAACCGCATCAGCACCACC...
```

Let us begin by checking the lengths of the two bacterial genomes in our example. They are stored in the files Mycobacterium_tuberculosis.txt, and Salmonella_enterica.txt (you can find these files on the website). The following commands will open the files, and read the genomes into two variables with the appropriate names:

```
1  f = open("./Mycobacterium_tuberculosis.txt")
2  Mycobacterium_tuberculosis = f.read().replace("\n","")
3  f.close()
4
```

```
5 f = open("./Salmonella_enterica.txt")
6 Salmonella_enterica = f.read().replace("\n","")
7 f.close()
```

Code explained

The function `open` takes a string representing the location of a file. Recall that ./ stands for the current directory (in which the Python script file containing these commands is located). The function `read` loads the whole content of the file. Since a text file may contain multiple lines, which end with the special 'end of line' character '\n', we need to get rid of this character, as it is not a part of the "real" genome. The `replace` method of class `str` does that.

Now, we can check the lengths of the genomes:

```
>>> n1 = len(Mycobacterium_tuberculosis)
>>> n1
4411532
>>> n2 = len(Salmonella_enterica)
>>> n2
4809037
```

We see that these genomes are several millions of nucleotides long, which is quite common for a bacterial genome.

Note: Don't try to print these genomes – they are too long, and most probably your IDLE will get stuck! But you can print a shorter slice of these genomes if you wish:

```
>>> Mycobacterium_tuberculosis[0:100]
'TTGACCGATGACCCCGGTTCAGGCTTCACCACAGTGTGGAACGCGGTCGTCTCCGAACTTAACGGC
GACCCTAAGGTTGACGACGGACCCAGCAGTGAT'
```

Both genomes contain, for example, the substring 'GA'. But do they contain longer, identical substrings? We will now devise several possible solutions to this problem and discuss their efficiency.

Exercise 1

'AGGGG' is also a common substring to both genomes.

(a) Find the positions (indices) where 'AGGGG' first appears in each of the genomes.
   Use the method `find` of class `str`.
(b) How many times does 'AGGGG' appear in each genome? Use the method `count` of class `str`.

3.1.2 Definition of the Problem

Before solving a computational problem, it is advisable to formalize it in terms of input and required output. This step is important because we are going to communicate with a computer in order to solve the problem, and thus we must be very specific and accurate. Here is the appropriate formalization for our case:

Input: two sequences, s_1 and s_2, and a positive integer k.
Output: a common substring of s_1 and s_2 of length k (if exists).

A substring of a given string is sometimes referred to as a "window" or a "block". In the biological context, and when the length of the substring (k) is a parameter, the term '**k-mer**' is very common. The output, in the case when a common substring of length k does not exist, should be some special value, which will indicate that this is the case. Python's None value is very convenient for this purpose.

When devising a solution to this problem, we should strive for an efficient one, in terms of running time and memory allocation requirements, as a function of the input size. In our case, the input contains three elements (s_1, s_2, and k), and we should consider how the size of each affects the complexity. We will denote the size of s_1 by n_1, and that of s_2 by n_2. On top of this, we should recall that other important criteria of any solution are their simplicity and readability.

3.1.3 A Naïve Solution

A **naïve solution** is a term describing an algorithm that solves a problem without any sophisticated techniques being employed. A naïve solution is usually simple to design and comprehend. In some cases, a naïve solution may mean searching for an appropriate output exhaustively. Algorithms of this type are also termed **brute-force** algorithms, or **exhaustive** algorithms. Sometimes, such algorithms are efficient enough, providing the result within reasonable time, especially if the problem size is not too large (in our case, sequences that are not very long for example). However, such solutions often turn out to be too slow for moderate or large size inputs. When this is the case, more sophisticated approaches should be sought. It is still a good idea to start by considering a naïve approach, even when it is suspected to be inefficient, just to get some idea about what we are trying to improve and possible ways to do that. Sometimes a naïve solution may give us intuition about the source of inefficiency, and therefore help direct us to better solutions.

Let us start by looking for a naïve solution to our common substring problem. The straightforward way to seek for a common substring of length k is simply to go over all possible pairs of k-mers: compare every length k substring of s_1 to every substring of s_2 of the same length.

```python
1 def common_substring_naive(s1, s2, k):
2     ''' find a common substring of length k in both s1 and s2 '''
3     for i in range(len(s1)-k+1):
4         for j in range(len(s2)-k+1):
5             if s1[i:i+k] == s2[j:j+k]:
6                 return s1[i:i+k] # return first match found
7     return None
```

> ### Code explained
>
> The function `common_substring_naive` checks all length `k` windows of the string `s1` vs. all those of `s2`. In line 3, we iterate over all starting positions of `s1`'s windows, which are the indices `0<=i<=len(s1)-k` (note that the rightmost index is `len(s1)-k`, since a substring of length `k` that starts at that index ends in the last index of `s1`, which is `len(s1)-1`). In line 4, we do the same for `s2`'s windows, and in line 5 we compare these two windows. Note that the function terminates immediately when it finds the first match in line 6 (and returns that match). If no such common window exists, it returns `None`.

A running example, with $k = 10$:

```
>>> common_substring_naive(Mycobacterium_tuberculosis,
    Salmonella_enterica, 10)
'TTGACCGATG'
```

> ### Exercise 2
>
> Change the function `common_substring_naive` such that it will return not only the matching k-mer, but also the indices in s_1 and s_2 where this k-mer starts. Let us call the new function `common_substring_naive2`. In formal terms, the new function should return also i, j such that $s_1[i : i + k] == s_2[j : j + k]$. For example,
>
> ```
> >>> common_substring_naive2(Mycobacterium_tuberculosis,
>     Salmonella_enterica, 10)
> ('TTGACCGATG', 0, 5940)
> ```

> ### Exercise 3
>
> Suppose we wanted to get *all the matches* of identical windows, not just the first match. Change the function `common_substring_naive` to return a list containing all the matches. Call the new version `common_substring_all_naive`. For example:
>
> ```
> >>> common_substring_all_naive("AGGAT", "GGACGAT", 3)
> ['GGA', 'GAT']
> >>> common_substring_all_naive("AGGAT", "GGACGAT", 4)
> []
> ```

A natural question is whether our naïve solution is good enough. In some cases, it is definitely sufficient. In the above running example for the two genomes and $k = 10$, we got a result instantaneously. However, the reason we got a result so fast was not because our algorithm is so efficient, but because a common substring was found reasonably early in both strings. Try running it with $k = 100$, for example:

```
>>> common_substring_naive(Mycobacterium_tuberculosis,
     Salmonella_enterica, 100)
...
```

This execution did not terminate after one minute, or even one hour, on our "standard" computer. What is the problem, then? At this point, a good idea will be to estimate how much time the execution for $k = 100$ roughly requires. We can do that by adding "diagnostic printouts." For example, we can print the index i in every iteration of the outer loop (see line 7):

```
1 def common_substring_naive(s1, s2, k):
2     ''' find a common substring of s1 and s2 of length k '''
3     for i in range(len(s1)-k+1):
4         for j in range(len(s2)-k+1):
5             if s1[i:i+k] == s2[j:j+k]:
6                 return s1[i:i+k]
7         print(i)                          # added this line!
8     return None
```

On a "reasonable" computer, for the same s_1, s_2, a new value of i is printed about every second (this may be different on your own computer, but it is not expected to be more than 10 times faster or 10 times slower). So how many seconds are needed for the execution to terminate? There are $n_1 - k + 1$ many is (this is exactly the number of iterations of the outer loop). Assuming a constant running time per iteration, the execution will terminate after 4411532-100+1 = 4411433 seconds. That is about *51 days!*

We should ask why the case of $k = 100$ is so much slower than $k = 10$. You may think that the larger k is, the longer the running time will be. As we will see later, this is not necessarily true (you may want to stop and think, why now). In addition, even if that were true, k that is 10 times larger would not explain this dramatic slowdown. The reason for the huge difference in running time here is completely different: for $k = 10$, our algorithm found a common k-mer. In fact, it found it rather quickly, in the first iteration of the outer loop, where i = 0, and for j = 5940 in the inner loop (as can be seen in Exercise 2). So in fact, the outer loop (where the running index is i) did not even have to get to its second iteration! A match was found after 5940 iterations of the inner loop, and the function terminated. So, for $k = 10$ our algorithm performed much less work than the worst possible case (and a bit more than the best case). Obviously, the faster the algorithm finds an identical substring, the faster it terminates.

For $k = 100$, the story is different, because as it turns out there is no common substring of length 100 for these genomes (although with the current naïve solution we would have to wait for about 51 days to realize that!). In terms of complexity, this is the worst case – the algorithm has to go through all the iterations to terminate.

> ### Exercise 4
>
> (a) What is the best case for the naïve solution?
> (b) In the worst case, how many iterations are needed in total (taking into account both loops)?
> (c) For much larger values of k beyond some point, the running time may actually decrease. Can you explain this?

3.1.4 Python's Sets to the Rescue

Now consider the following alternative solution:

```python
1   def common_substring_hash(s1, s2, k):
2       ''' find a common length k substring of s1 and s2
3           using Python built-in sets '''
4       table = set()
5       for i in range(len(s1)-k+1):
6           table.add(s1[i:i+k])
7
8       for i in range(len(s2)-k+1):
9           if s2[i:i+k] in table:
10              return s2[i:i+k]
11
12      return None
```

> ### Code explained
>
> In line 4, we initialize an empty set. Recall that sets are unordered collections of elements. Choosing sets, rather than for example lists, is crucial, as we will explain later. The loop in lines 5 and 6 inserts into the set all the k-mers of s1, using the method add of the class set. Then the loop in lines 8–10 checks, for each of the k-mers of s2, if it appears in the set, and returns the first match found, or None if no match is found.

Let's put this solution to the test of time:

```
>>> common_substring_hash(Mycobacterium_tuberculosis,
    Salmonella_enterica, 100)
>>>              <= terminated and returned None
```

The function returned None (as mentioned earlier, there is no common substring of length 100 in these two genomes). But the main point here is that it terminated after just a few seconds! How come it performs so much faster than the naïve solution? What's the magic? The short answer is that Python's sets are implemented as hash tables, which is the topic of this chapter. This surprising improvement in running time certainly deserves some elaboration.

3.2 Hash Functions and Hash Tables

3.2.1 Hash Functions

Consider the following function:

```
1 def hash4strings(st):
2     s = 0
3     for c in st:
4         s = (128*s + ord(c)) % (2**120+451)
5     return s**2 % (2**120+451)
```

This weird looking function takes a string as its input and returns an integer that is computed from that string. Python's built-in function `ord` is used in line 6. This function returns the Unicode value of a given character. Without getting into all the details, each character has an integer representing it. For example, "a" is 97, "A" is 65 and " " (space) is 32:

```
>>> ord("a")
97
>>> ord("A")
65
>>> ord(" ")
32
```

`hash4strings(st)` does some calculation on the characters of the input string one by one. The details are not that crucial. What is important here is the fact that this function maps any string to a number. Specifically, it maps infinitely many strings to a large, yet bounded, range of integers (between 0 and $2**120 + 450$. Do not be bothered by this specific number, as it bears no special meaning in our current context).

A few running examples:

```
>>> hash4strings("A")
4225
>>> hash4strings("ATTA")
18961827860014209
>>> hash4strings("GTTA")
22585540648068225
```

This is an example for a **hash function**, which maps strings to numbers in a bounded range. More generally, a hash function maps a large (possible infinite) set of elements into a much smaller (and finite) range of integers. You may wonder how such a function can be applied in the context of our original problem, and this will indeed be clarified very soon.

Python has its own built-in hash function, appropriately named `hash`. This function can be applied to strings, as well as numbers and objects of additional types (but not to all types). Here are some examples: (Important note: the outcomes vary from one machine to another, and even from one IDLE session to another, because some randomization is used behind the scenes. So running these and later examples will most probably yield different results.)

```
>>> hash(3)
3
>>> hash(3**100)
1175369268131054105
>>> hash(5.9)
2075258708292325381
>>> hash("A")
-9843664961898564554
>>> hash("ATTA")
5452722747407986471
>>> hash("GTTA")
-3679160727658122423
```

The values produced by Python's hash belong to a large range of integers, which actually include negative integers as well. We are not particularly interested in the inner details of Python's hash. All we care about here is the basic fact that this function maps infinitely many elements into a fixed, finite range.

Another similarity between our and Python's hash functions is that both perform the mapping in a way that looks unexpected, and even "similar" objects are mapped to completely different values. For example, changing ATTA to GTTA greatly changed the result. Put differently, it seems that this function performs a "messy" mapping, which looks random (although it is not).

Lastly, note that since a hash function maps a large universe of elements to a smaller one, the hash function cannot be **one to one** – there must be different elements that are mapped to the same number. Such a scenario is known as a **collision**. Collisions are unavoidable in our context, due to the well-known "**pigeonhole principle**" – when n pigeons are placed into m holes, and $n > m$, there will surely be a hole with at least two pigeons. We will get back to this crucial point later. For now, let's assume the collection of elements we hash is small enough that no collisions occur.

3.2.2 Hash Tables

Hash functions are a very powerful approach for storing elements for a later lookup. To understand how, first recall that you can store elements in a list, simply by appending new elements to the end of the list. However, when looking up a specific element, you may well go over a large portion of the list to find the element, or even go through all the list, when the element is not in the list. Hash functions allow us to use the allocated memory of the list more cleverly. Let's see how.

Suppose we have a table with m entries, which you can use to store your elements. When storing an element, say, a string, st, we can store it at index $i = $ hash$(st)\%m$. Note that this is an integer between 0 and $m - 1$. For example, if $m = 7$, and suppose hash(st) is 5452722747407986471, then hash$(st)\%7$ is 1, so st will be stored in the second cell of the table (at index 1).

More generally, an element x will be stored at index $i = $ hash$(x)\%m$ of a list of size m. We call that list a **hash table**, and denote it by T. From now on, we will simply denote hash$(x)\%m$ by $h(x)$. The function h maps elements to integers in the range of 0 to $m - 1$

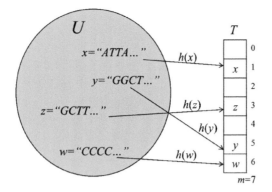

Figure 3.1. A large universe U mapped into a hash table T with seven slots using a hash function h

(formally, $h: U \rightarrow \{0, 1, \ldots, m\text{-}1\}$). We view these numbers as indices in a table of size m (see Figure 3.1).

The crucial point is that when you look up some item x in the table, you look for it at index $h(x)$ only, instead of traversing the whole list! In terms of our common substring problem, we can keep all the k-mers of s_1 in a hash table, and then look up each k-mer of s_2 in the table very efficiently. We avoid comparing every pair of k-mers from the two strings. This is a huge saving!

Exercise 5

In our original common substring example, the universe of elements was the set of all possible genomic k-mers for a given k.

(a) What is the number of elements in this universe (that is, how many different k-mers are possible, over the DNA alphabet $\{A,T,C,G\}$, as a function of k)?

(b) How many different k-mers can a string of length n contain?

(c) Consider the genome of *Salmonella enterica*, whose length is $n_1 = 4809037$, and suppose $k = 100$. We want to store all 100-mers of that genome. If all the 100-mers in this genome are different, what portion (fraction) of the universe are we expected to store?

It is time to confront the problematic issue of collisions. Since collisions are unavoidable, two different elements may be mapped into the same index in a hash table. This puts our "magical approach" at risk. Furthermore, not only two different elements may collide under the original function hash(): when we narrow down its range even more, by computing it modulo m (to fit the indices of a table of size m), elements that did not collide under hash() may actually collide under $h(x) = \text{hash}(x)\%m$. For example:

```
>>> hash("Benny")
2110211825
>>> hash("ATTA")
5452722747407986471
>>> hash("ATTA")%7 == hash("Benny")%7
True       <= both %7 are 1
```

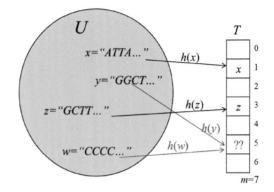

Figure 3.2. Collision of two elements

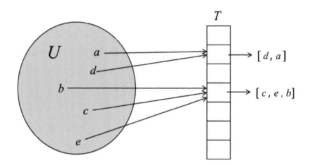

Figure 3.3. Chaining to resolve collisions

Generally, a collision is a scenario in which $h(x) = h(y)$ for $x \neq y$, as illustrated in Figure 3.2.

A good hash function, one that spreads the elements fairly evenly, will minimize the number of collisions. However, one certainly cannot avoid all collisions.

3.2.3 Dealing with Collisions

There are several approaches to dealing with collisions. We will describe two common ones. Perhaps the simplest approach to think of is **chaining**: keep a separate list that is "linked out" from every cell (index) of the table. These lists are termed chains. All elements that are mapped into a specific index of the table will be contained in the chain of that index (see Figure 3.3).

Here is the implementation of the standard operations in a hash table with chaining:

- Insert(x) – append x to the chain at index $h(x)$ (initially, the chain is empty).
- Search(x) – go over the chain at index $h(x)$ and look for x there. If the search failed, x is not stored in the table.
- Delete(x) – search for x. If it is found, remove it. If not, do nothing.

Exercise 6

Suppose 63 ID numbers are inserted into a hash table of size $m = 10$, with the hash function hash(id) = $id\%10$. That is, the hash function value is the rightmost digit of the ID number.

(a) What is the average length of the chains?
(b) Assume the hash function distributes the elements among the table indices "optimally." Would you expect one or more chains of length 6? How about 20?
(c) Can you provide an example for a worst-case set of IDs for which this hash function creates one chain of length 63 while keeping all the other chains empty?

Armed with a good hash function that distributes the elements reasonably uniformly in the table, one would expect the longest chain not to be much longer than the *average* chain length. If we store n elements in a table of size m, using chaining to resolve collisions, then the average number of elements in any chain (the chain length) is n/m. The punch line is this: if we choose m such that it is equal or close to n, then we have chains of average length 1 or close to 1. This means that, on average, when looking up an element in the table, we will have, on average, to check $n/m \approx 1$ element.

One should not forget that in the worst case, things are different. In the worst-case scenario, all the elements are mapped into the same chain. This resulting hash table contains one chain of length n, while all the other chains are empty. Searching an element in the table may result in going over all n elements. This is no better than simply storing the elements in a list. However, this worst case is extremely rare, and the probability of it occurring (probability over the collection of n elements) is extremely small. This is why, in practice, hash tables enable a very efficient search, with a constant number of operations (more generally, close to n/m).

There is another approach to dealing with collisions, called **open addressing**. This is in fact the approach used by Python's sets and dictionaries. Very briefly, in this approach each cell in the table contains at most one element. When a new element tries to enter an already occupied cell, either the new element or the old one will be moved to another cell, according to some predefined rule. For example, the new element can simply look for the next empty cell in the table by advancing forward one cell at a time (see Figure 3.4).

This may raise some issues, such as how lookup is performed in the open addressing approach, what average number of cells need to be visited when searching an element, etc. We will not go into these, but merely mention that, in practice, this approach is at least as efficient as chaining, and in some cases even more efficient.

Python's sets and dictionaries allow efficient lookup because they are implemented as hash tables (in dictionaries, it is the keys that can be searched for efficiency, not the accompanied values). The hash function used is naturally Python's own `hash`. Collisions are resolved using a rather sophisticated implementation of open addressing. Normally, all the inner workings of these are transparent to us. The exact details may even slightly change between Python versions. All we need is to know the interface for using Python's sets and dictionaries as "black boxes" and enjoy the fact that they are so efficient. How very convenient!

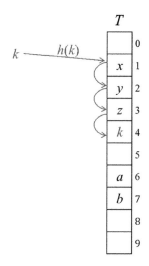

Figure 3.4. Open addressing example to resolve collisions

Box 3.1 - A Detour: The birthday paradox

Recall that if the universe is larger than the table size, collisions in a hash table are unavoidable. What's more, the expected number of collisions grows when the number of elements in the table increases. When the ratio n/m is large, the table is more loaded, and the expected number of collisions is higher. The ratio n/m is called the **load factor** of the table.

An interesting phenomenon, known as the "birthday paradox," is closely related to the collision probability. Suppose you are in a (fairly large) room, with 23 people (including yourself). What is the probability that two of them (including yourself) have the same birthday (day and month. It is impolite to ask about the year these days, and in some settings it is even illegal, as in interviews for job application)? The probability here is over the birthdays of the people in the room.

It turns out that the probability for such collision is 0.5073, i.e., strictly greater than 0.5. This contradicts "common intuition," which assigns a much lower probability to this event. If there are 30 people in the room, the birthday collision probability goes up to 0.7063.

How is this cool "paradox" related to hash tables? The "objects" are the people in the room. The hash value of a person is his/her birthday. There are 365 possible days, which is the size of our imagined hash table. It turns out, that when hashing 23 objects into a table with 365 entries, the probability of at least one collision is 0.5073.

Note that we assume that the people in the room got there independently of their birthday. If you handpicked 30 people, born on the 30 different days of April, there will obviously be no collisions.

3.2.4 Time Complexity of the Hash-Based Solution

Back to the function `common_substring_hash`, the general structure of the code is the following:

---

common_substring_hash(s_1, s_2, k)

1. Create an empty hash table of size m
2. Insert all k-mers of s_1 into hash table (the address is hash(string) % m)
3. For every k-mer of s_2,
    3.1 if it is in the hash table – return it and finish
4. Declare no common substring of length k

We can now understand why this solution is much more efficient than the naïve one. Specifically, consider the worst-case input for the previous solution, which is when no common k-mer exists. The total number of iterations in the naïve solution was $(n_1 - k + 1) \cdot (n_2 - k + 1)$. In the hash-based solution, we execute only $(n_1 - k + 1) + (n_2 - k + 1) = n_1 + n_2 - 2k - 2$ iterations. In the naïve solution, the number of iterations is the product of the two terms, whereas in the hash-based solution we have the sum of the same terms. Take $k = 100, n_1 = 4{,}411{,}532$ and $n_2 = 4{,}809{,}037$, for example. The naïve solution takes about $2 \cdot 10^{13}$ iterations, whereas the hash-based solution takes about 10^7 iterations, a multiplicative speedup of two million! Note that each iteration in the naïve solution required comparing two k-mers, while in the hash-based solution we compare at least two k-mers per iteration on average, and may require more comparisons. This depends on the collisions, and the inner workings of the hash table implementation, its size, etc. But the bottom line remains the same – in practice, we get a major improvement, achieved by storing elements in a hash table for efficient lookup.

3.3 Improving Memory Requirements

3.3.1 Shorter Is Better!

A major difference between the naïve solution and our hash-based solution is that in order to save on running time, the latter algorithm uses the computer's memory to store all the k-mers of s_1 in a table. More precisely, it needed to store exactly $n_1 - k + 1$ substrings, each of length k. In total, $k \cdot (n_1 - k + 1)$ characters are stored. When n_1 is a few millions, and k is not very small (say, 1000 characters long), this has implications for the ability of a reasonable computer to execute the algorithm without getting stuck, due to a **memory overload**. If a single character requires 1 byte (8 bits), for example, this solution requires more than a billion bytes (1 Gigabyte) of memory to run. If n_1 is a few tens of millions, the memory requirement becomes 10 Gigabytes, which may already be too taxing on your personal machine. Insufficient free memory will result in a memory failure error, and IDLE will simply crash! (Do not try, for example, to run it on $k = 1000$. Well, you might want to try, just out of curiosity, but hey – don't say we didn't warn you.)

Can we do anything about it? When n_2 is much smaller than n_1, we surely can by switching roles between s_1 and s_2. In other words, we use the shorter string for hashing, and the longer one for searching. We can do that at the level of the function call, i.e., compare the lengths of the two strings before calling the function:

```
>>> if len(s1) < len(s2):
        common_substring_hash(s1,s2,k)
    else:
        common_substring_hash(s2,s1,k)
```

But there is a nicer way to do it, which hides this issue inside the function itself:

```
1   def common_substring_hash2(s1, s2, k):
2       ''' find a common length k substring of s1 and s2
3           using python built-in sets '''
4       if len(s2) < len(s1):
5           s1,s2 = s2,s1     # simultaneous assignment
6
7       table = set()
8       for i in range(len(s1)-k+1):
9           table.add(s1[i:i+k])
10
11      for i in range(len(s2)-k+1):
12          if s2[i:i+k] in table:
13              return s2[i:i+k]
14
15      return None
```

🔍 **Code explained**

Lines 4 and 5 are the only addition to the previous version. We simply switch `s1` and `s2`, in case `s2` is shorter. This way, when we proceed we can be sure `s1` is not longer than `s2`!

When s_2 is much shorter than s_1, we expect a significant improvement in memory usage. Consider a third genome: *Mycobacterium leprae*, which is 3,268,071 nucleotides long, shorter than its relative *Mycobacterium tuberculosis*, which is of length 4,411,532. We first run the original version of our hash-based algorithm on these genomes with $k = 150$, then try the memory saving version. We remark that in both executions, the shorter genome of *Mycobacterium leprae* is provided the second parameter s_2 (why?). Indeed, memory usage inspection shows a difference, as you can see in Figure 3.5.

```
>> common_substring_hash2(Mycobacterium_tuberculosis,
    Mycobacterium_leprae, 150)
'TCTCGGATTGACGGTAGGTGGAGAAGAAGCACCGGCCAACTACGTGCCAGCAGCCGCGGT
AATACGTAGGGTGCGAGCGTTGTCCGGAATTACTGGGCGTAAAGAGCTCGTAGGTGGTTTG
TCGCGTTGTTCGTGAAATCTCACGGCTTA'
```

This issue may seem of little significance to you, but these differences could be crucial when pushing the size of the problem to the limit of a specific computer in terms of available

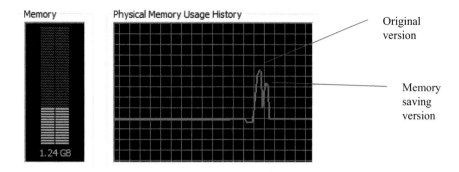

Figure 3.5. Memory consumption differences when constructing the hash table on the longer and shorter genome

memory. For example, the computer, on which these examples were run, could not handle $k = 200$ for the original version (it crashed due to a memory overflow!), but did terminate successfully with the memory saving version.

In fact, the memory saving version improves not just the memory, but the running time as well, due to the cost of handling large structures in the computer's memory. The overhead of checking the lengths of the genomes and switching between them is negligible in terms of running time, while the improvement in the main body of the execution may be significant. Specifically, we observed a speedup of about 20% in running time.

Are we done with memory improvements? Certainly not! First, recall that when the strings are of similar lengths, the memory saving is negligible. Therefore, we expect the two versions to perform about the same in terms of memory (and time) consumption. Furthermore, as mentioned earlier, when k is more than a few hundreds, and for genomes of a few million bases, even the memory saving version fails due to a memory overflow (on any "reasonable" home computer).

When the lengths of the biological sequences are in the scope of millions, our current solution is good for finding common substrings of length up to a few hundreds. In many biological contexts, this is not enough. In the next section, we introduce an important memory saver method – fingerprints.

3.3.2 Fingerprints

In our solution, we store many substrings in the table. The exact number is the minimum between $n_1 - k + 1$ and $n_2 - k + 1$. We cannot reduce this number, since if we skip some of the substrings we may get an incorrect, false negative result. However, we store k characters for each such substring, and we can reduce this quantity. Suppose we could represent a string in a compact way, and store that representation in the table, instead of the string itself. This would save a lot of memory. Such a compact representation is called a **fingerprint** (just like human fingerprints, used for identification). Our improved algorithm will insert into the table their fingerprints rather than the k-mers and search for repetitions. But first, we should figure out: what is that 'compact representation' for strings?

We could take, for example, the first character of the string as a (very) compact representation. Instead of k characters, the table will consist of just one character for each substring. However, this is a bad idea, since many substrings begin with the same character, and their representations will be identical. For genomic strings, we will have only four such different representations, and we will not be able to detect almost any of the repetitions. What if we took the first 10 or 100 characters? This is better, but still suffers from the fact that only a specific part of the substring is used to represent the substring. It would be more reasonable that all the characters affect the compact representation, so even if two strings are only one character apart their representations will differ with high probability.

In fact, we have already seen such a compact representation method. Take for example the function `hash4strings` we saw earlier. It transforms a given string into a number, in the range 0 to 2**(120 + 450). We could use this number as the fingerprint of the substring. Actually, we could just use Python's `hash` function instead of our `hash4strings`!

This may be a bit confusing: Python's hash now has two roles in this story. It is used to map elements into indices of a hash table, and also to generate a compact representation of long strings in order to save memory. When we use Python's sets as our hash tables, the first role is done "behind the scenes."

An inherent issue that arises is that, like before, there may be collisions. This time, we mean two strings that have the same fingerprint. This is generally unavoidable, because we transform the strings into elements in a smaller space, and some of the information is lost by such transformation. For this reason, when we encounter a fingerprint that has already been inserted into the table, this does not necessarily mean that we have identical substrings. Such a situation indicates a "suspect" for a match, as we then need to go and check the original substrings themselves. The bottom line is that in most cases, we will get a negative feedback at the fingerprint level and will not have to look at the original substrings.

```python
1   def common_substring_fingerprint(s1, s2, k):
2       ''' find a common length k substring of s1 and s2
3           using python built-in sets and fingerprints to save memory '''
4       if len(s2) < len(s1):
5           s1,s2 = s2,s1
6
7       table = set()
8
9       for i in range(len(s1)-k+1):
10          fingerprint = hash(s1[i:i+k])
11          table.add(fingerprint)
12
13      for i in range(len(s2)-k+1):
14          fingerprint = hash(s2[i:i+k])
15          if fingerprint in table: # possible match
16              if s2[i:i+k] in s1:    # sanity check
17                  return s2[i:i+k]
18              else:
19                  print("ALMOST A FALSE POSITIVE:", s2[i:i+k])
20      return None
```

Code explained

The difference between this function and the previous one is the use of fingerprints. In lines 10 and 11, instead of inserting the whole substring `s1[i:i+k]` into the hash table, we insert its fingerprint `hash(s1[i:i+k])`. This is supposed to save a lot of memory. The same happens in the second loop: in lines 14 and 15, we search for the fingerprint of `s2[i:i+k]` in the table (instead of the substring itself). However, if we find that fingerprint in the hash table, we still need to make sure it represents a true match, and not just a match in fingerprints of two unequal strings. This verification is done in line 16, which checks if the current substring of `s2`, `s2[i:i+k]` exists in `s1`.

If you were curious, the following command returned `None`, meaning these two genomes have no common substring of length 1000. No memory failure this time!

```
>>> common_substring_fingerprint(Mycobacterium_tuberculosis,
    Salmonella_enterica, 1000)
>>>                    <= None returned
```

3.4 Dictionaries and the Most Frequent *k*-mer Problem

The usage of Python's sets is somewhat limited: they only indicate whether an item is present or not. Sometimes, we need to attach additional information to these items (not just "present" or "absent") for some purpose. For example, if we deal with items that are substrings of a genome, we may want to know the number of times each substring appears in the genome, its starting position, etc. This is where Python's dictionaries (type `dict`) step in.

In this section, we will show how Python's dictionaries can be employed in order to study various *k*-mer statistics. Specifically, we will study the problem of finding the **most frequent *k*-mer** in a genome.

Suppose we wanted to find the most frequent substring of a given length in a given string. That is, for a string *st* and a length *k*, we want the substring that appears the largest number of times in *st*:

Input: a string *st*
         an integer *k*
Output: a *k*-mer contained in *st* the maximal number of times.

Storing the substrings of *st* in a set will not produce this information. We also need to count how many times each *k*-mer appears in *st*. In other words, we will attach to each substring of *st* a counter, representing its abundance. Then, we can take the one with the maximal count as the output. So what we need is to map between substrings of *st* and their counters.

To store both the substrings and their counts, we will use Python dictionaries. Python's dictionaries are implemented, behind the scenes, as hash tables as well. Thus, searching for an element in them is very efficient. For example, for this input:

```
st = "CCCTTGCTT", k = 3
```

we will build the following dictionary:

```
{'GCT': 1, 'TTG': 1, 'CTT': 2, 'TGC': 1, 'CCT': 1, 'CCC': 1}
```

The most frequent *k*-mer here is 'CCT', which appears twice in *st*. So, an efficient solution for the frequent repeating substring problem may look like this:

```
1   def frequent_kmer(st, k):
2       counters = dict()
3       most_freq = ""
4       max_freq = 0
5
6       for i in range(len(st)-k+1):
7           if st[i:i+k] not in counters:
8               counters[st[i:i+k]] = 1   # first time
9           else:
10              counters[st[i:i+k]] += 1 # found one more
11              if max_freq < counters[st[i:i+k]] :
12                  max_freq = counters[st[i:i+k]]
13                  most_freq = st[i:i+k]
14
15      return most_freq, max_freq
```

Code explained

Line 2 initializes an empty dictionary, to be used to store k-mers and the corresponding counters. Lines 3 and 4 initialize the variables that will store the current results, along with the execution of the algorithm. In the loop that starts at line 6, we iterate over all the k-mers of st. If the k-mer was encountered for the first time (line 7), it is inserted into the dictionary with counter = 1 (line 8). Otherwise, its counter is increased by 1 (line 10), and the temporary result variables are updated if needed (lines 11–13).

```
>>> frequent_kmer(Mycobacterium_leprae, 3)
('CGG', 89360)
>>> Mycobacterium_leprae.count('CGG')    <= sanity check using Python built in method
89360

>>> frequent_repeating_substring(Mycobacterium_leprae, 200)
('GTCATGGCCTTGAGGTGTCGGCGTGGTCAATGTGGCCGCACCTGAACAGGCACGTCCCCGTGCACGGTATAA
CTATTCGCACCTGATGTTATCCCTTGCACCATTTCTGCCGCTGGTATCGGTGTCGGCGGCTTGTTGACCGGCCC
TCAGCCAGCAAGCAGGCATGCCGCCGGGTGCAGCAGTATCGTGTTAGTGAACAG', 19)

>>> frequent_repeating_substring(Mycobacterium_leprae, 400)
('TGCTGCTTGGTCTACATGTTGATGATGCCAGGGGCTGGGCACCTGGGCTGTGCTGAAGGCGATATCGATGCA
GGCGTGAGTGTGAGGATAGTTGTTAGCGCCGCGGGGTAGGGGCGTTTTAGTGTGCATGTCATGGCCTTGAGGTG
TCGGCGTGGTCAATGTGGCCGCACCTGAACAGGCACGTCCCCGTGCACGGTATAACTATTCGCACCTGATGTTA
TCCCTTGCACCATTTCTGCCGCTGGTATCGGTGTCGGCGGCTTGTTGACCGGCCCTCAGCCAGCAAGCAGGCAT
GCCGCCGGGTGCAGCAGTATCGTGTTAGTGAACAGTGCATCGATGATCCGGCCGTCGGCGGCACATACGGCAAC
CTTCTAGCGCAGATCAACCACCCACACCCCAC', 9)
```

For larger k' s, this solution encountered the same memory problems as before. This solution could be improved using the mechanism of fingerprints to save memory, but in this case it is more complicated, and we will not show it.

Reflection

In this chapter, we introduced a fundamental concept in computer science – hashing. There are two different notions related to hashing: hash functions and hash tables. A hash function, is a mapping from a large set of elements, called a universe, to a smaller set of numbers or memory addresses. This enables us to store elements, e.g., strings, in a table, called a hash table. A good hash function spreads the elements well in the table, thus minimizing the probability of a collision.

The essence of using hash tables is that we search within a limited set of candidates. So, instead of searching through all elements (as in a "regular" list for example), hash tables separate the elements into "bins," thus narrowing down future searches to just elements in the appropriate bin. As we said, given a good hash function and appropriate table size, the chains are expected to be short.

Also, hash functions may be used without subsequently inserting elements into a hash table. For example, in this chapter we used the hash function as *fingerprints*: a compact representation of the element being hashed. This allowed us to save a lot of memory, allowing our solution to be applicable for longer strings.

Hash tables are one (important) example of what computer scientists call a **data structure**. This is a way to organize data in the computer's memory, in such a way that will enable efficient operations. Some programming languages provide built-in, ready-to-use data structures, such as Python's lists and Python's sets and dictionaries. Other, more sophisticated data structures may not be provided as part of a language and require specific implementation for a specific use. The choice of the appropriate data structure depends on the required operations and efficiency.

In our example, hash tables enable very efficient table insertion and lookup. Other applications will typically require a different data structure. For example, if one wants to dynamically store and delete strings, and find the longest string at any given time, a hash table would not be the right choice. Finding the longest string in a hash table would require examining the whole table. A simple list sorted by the length of the strings would be more appropriate in this case: at any moment, the longest string would be located at the end of the list. However, this requires maintaining the sorted order of the list at every insertion of a new string. There are more sophisticated data structures that balance this tradeoff in different ways, but this is a topic for other books. Later in this book we will, however, meet additional data structures.

Challenge Yourself

Problem 1 The common substring problem using Python's 'in' operator

Look at the following solution to the common substring problem, which uses Python 'in' operator:

```
1 def common_substring_better(s1, s2, k):
2       ''' find a common substring of s1 and s2 of length k '''
3       for i in range(len(s1)-k+1):
4           if s1[i:i+k] in s2:
5               return s1[i:i+k]
6       return None
```

Code explained

Line 4 simply checks if the substring `s1[i:i+k]` is a substring of `s2`, using Python's `in` operator. The value of the expression in this condition is either `True` or `False`. All the rest remained the same, but note that we now have only one explicit loop. The inner loop from the naïve solution is hidden under Python implementation of the `in` operator.

Very often, using Python's built-in functions, operators, or structures shortens and simplifies the code. In this case, using the 'in' operator replaced an explicit loop. However, we should remember that a simpler or shorter code does not always imply a more efficient one. In particular, the 'in' operator still involves iterating over s_2 in some manner.

It turns out Python's 'in' operator is implemented very efficiently. As in many other built-in functions and operators, Python's developers make an effort to optimize efficiency. The details of this implementation are not simple at all and involve rather sophisticated algorithms (based on a mix between Boyer–Moore and Horspool algorithms, which are way beyond the scope of this book). So, in fact, this is not a naïve solution at all, if you take into consideration Python's implementation "under the hood." However, in the eyes of a programmer that treats the 'in' operator as a black box, the structure of this solution is certainly simple.

Provide an estimation of the running time for the following execution with $k = 100$:

```
>>> common_substring_better(Mycobacterium_tuberculosis,
    Salmonella_enterica, 100)
```

Add, as we did before, diagnostic printouts of the value of the variable `i`. You may want to print `i` not in every iteration, but every, say, 100 iterations. You could use the modulo (%) operator for that. How does the expected running time compare to the 51 days needed for the naïve version?

Problem 2 The longest common substring problem

In this chapter, we tackled the problem of finding an identical substring of a given length k in two strings. Moving a step forward, can we find the *longest* identical substring of two genomes efficiently?

Consider for example the genomes of *Mycobacterium tuberculosis* and *Salmonella enterica*. Would you expect these genomes to share an identical region of length 10? 100? 1000? More? What about *Mycobacterium leprae*, a closer relative of *Mycobacterium tuberculosis*? Can we find evidence for the evolutionary similarity between the two *Mycobacterium* in the form of longer identical regions?

Even with a simple "toy" example, the problem of finding the longest common substring is not easy to solve manually. For example, try to find the longest common substring of these two short sequences (how much time do you need?):

s_1 = "GATTAGCCGTAGATTGA"
s_2 = "AGGAAGGATGCCGTGAAA"

Yet, devising an algorithm to solve the problem is easy – at least the naïve algorithm.

(a) Solve this problem using `common_substring_hash2` on increasing values of k (1, 2,...).
(b) Run your function on s_1 and s_2 that appear above ("toy" example). What is their longest common substring?
(c) Try running your function on *Mycobacterium tuberculosis* and *Salmonella enterica*. How much time did it take?

One possible time saving approach is going over the values of k in a non-linear fashion (not going through k successively). We can use the binary search approach and check only a much smaller number of potential lengths. We can do the following: check a length in the middle between 1 and the maximum possible length (which is the minimum between the length of s_1 and s_2). If there is no match, we know that none of the larger lengths will yield a match either. This way we can keep on searching the range of lengths smaller than the middle. Otherwise (if there was a match), we should keep searching larger lengths, avoiding the time spent on smaller lengths. This approach would check logarithmically many lengths in the range 1 to the minimum of the strings' lengths.

However, this approach turns out to be problematic. Often, s_1 and s_2 are very long compared to the length of the longest common substring. In our case for example, these strings are millions of nucleotides long, while, as we will see soon, the longest common substring has less than 100 nucleotides. Had we followed the above-mentioned approach, we would have started checking a k that is around 2 million. Don't try that! It is more than your personal, home, computer is capable of computing.

So what else can be done to still be able to check lengths that are closer to the real maximal length of a common substring? We will still use the binary search approach, but in a slightly different manner. Consider this code:

```
1 def longest_common_substring_bsearch(s1, s2):
2     ''' using binary search on the length '''
3     print("Starting binary search on length of common substring...")
4     k = 1
5     while common_substring_hash4(s1,s2,k) != None:
6         print(k, "found")
7         k *= 2
8
9     print(k, "not found")
```

What we have done here, is first check increasing values of k, but each time we double k: k = 1, 2, 4, 8,... and so on. This resembles binary search, where instead of taking the middle each time, we started from the minimal value possible, and jumped ahead in multiples of 2 each time.

(d) Run this function on *Mycobacterium tuberculosis* and *Salmonella enterica*. In which range does the length of the longest common substring lie?

Once we have narrowed down the search space so dramatically, from a matter of millions to just a few dozen, we can go on and search the limited range, either linearly – or better yet – employing binary search again on the limited range, as in the code above. Here is the rest of the function:

```
1 def longest_common_substring_bsearch(s1, s2):
2     ''' using binary search on the length '''
3     print("Starting binary search on length of common substring...")
4     k = 1
5     while common_substring_hash2(s1,s2,k) != None:
6         print(k, "found")
7         k *= 2
8
9     print(k, "not found")
10
11    print("\nSearching for lengths between", k//2+1, "and", k-1, "...")
12    longest = k//2
13    low = k//2+1
14    high = k-1
15    while low <= high:
16        mid = (low+high)//2
17        res = common_substring_hash2(s1,s2,mid)
18        if res == None:
19            print(mid, "not found")
20            high = mid-1
21        else:
22            print(mid, "found")
23            low = mid+1
24            longest = mid
25
26    longest_ss = common_substring_hash2(s1,s2,longest)
27    print("Longest common substring of length", longest,":", longest_ss)
```

(e) Run the full version of the function on the two genomes. How much time did it take to find the result? What is it? How many *k*'s were checked?

(f) In the first stage of the algorithm (lines 3–9), instead of multiplying *k* by 2 in each iteration, we could multiply it by 3. Would that reduce the number of lengths the algorithm would inspect? Consider the case where the longest substring is of length 62.

(g) Find the longest common substring for each pair of genomes (*Mycobacterium tuberculosis*, *Salmonella enterica*, and *Mycobacterium leprae*). What is your conclusion?

Problem 3 Common substring for three strings

Write a function `common_substring_3(st1, st2, st3, k)` that returns a common substring of all three strings of length *k*. Run it on the three bacterial genomes. Try to be as time efficient as possible. What is the result?

4 Biological Patterns and Regular Expressions

In this chapter, we study another common string-related problem – pattern matching. Suppose we want to find a given sequence motif, or pattern, in a genome or protein, where the pattern is not unique. In other words, the pattern has more than a single possible matching sequence. To that end, we will introduce the fundamental notion of **regular expressions**, and their use to solve this problem. In addition, we will discuss the closely related notion of **finite state machines** (FSM), another basic concept in computer science.

4.1 Pattern Matching

The **exact string searching problem** is defined as follows:

Input: a string p of length m (called pattern), and another string s of length n (called text), where $m \leq n$.
Output: does s contain p as a substring?

We can think of several possible solutions to this problem.

First, we could use a naïve approach, which simply compares the pattern to every substring of s of length m. This solution works in time complexity of $O(nm)$, since each iteration takes $O(m)$ time in the worst case (comparing m characters), there are $n - m + 1 = O(n)$ such substrings of s, and therefore $O(nm)$ iterations in total.

A second possible solution would make use of hash tables: insert every substring of s of length m into a hash table (e.g., Python's set), and then check if p is in the table. In fact, this solution is no better than the naïve one, since inserting $O(n)$ substrings of length m each into the hash table already costs $O(nm)$ time. Furthermore, this solution requires more memory resources, since $O(nm)$ characters are stored in the hash table. Still, there is a scenario in which such a hash-based solution is justified in terms of efficiency. If one plans to search many different patterns of the same length m in the same string s, then storing the substrings of s in a hash table needs to be done only once. Then, each query of the form "is p in the hash table" costs $O(m)$ time. In other words, s was "pre-processed" for efficient future queries.

But now there's a twist, which complicates things. Suppose the pattern p is not merely a single string, but rather a general template, which fits many possible strings. We call this the **pattern matching problem** or the **motif search problem**. Indeed, many biological motifs allow some "degrees of freedom." For example:

- *'TATAA' followed by any three nucleotides and ending with 'TT'* (**TATA-box**)
- The amino acid *Asn, followed by any amino acid but Pro, followed by either Ser or Thr, followed by any amino acid but Pro* (**N-glycosylation site motif**)
- *'ATG' followed by 3n nucleotides (for some non-negative integer n), then 'TAG' / 'TAA' / 'TGA'* (**ORF, open reading frame**)

For these rather simple patterns, we can devise specific solutions, considering the alternative matching options. For example, to find a TATA-box in a Python string `s` we need to search for an `i` such that `s[i:i+5]=="TATAA"` and `s[i+5] in "ATCG"` and `s[i+6] in "ATCG"` and `s[i+7] in "ATCG"` and `s[i+8:i+9]=="TT"`. Even for this simple pattern, the solution is a bit cumbersome. When the patterns are more complicated, such an approach becomes very inconvenient, or even very inefficient.

4.2 Regular Expressions

Regular expressions (RE) are a convenient and popular way of representing textual patterns. They are expressive enough to capture many patterns we are interested in. Furthermore, they do so in a compact manner, as we will see. Regular expressions are an important component of the theory of computing and are closely related to the so-called deterministic finite automata discussed later. In our context, Python comes with a built-in package for regular expressions, which we will use throughout this chapter.

4.2.1 Python's `re` Module

Python has a built-in module, called `re`, for regular expressions. To invoke it, add `import re` at the beginning of your code.

Let's start with the following example:

```
>>> import re

>>> match = re.search("hello", "Hello dogs, hello cats!")
>>> match.group()      <= why "group"? we'll explain later
'hello'
>>> match.start()
12
>>> match.end()
17
>>> match.span()
(12, 17)

>>> match = re.search("Goodbye", "Hello dogs, hello cats!")
>>> match
>>>    <= None returned

>>> help(re.search)
Help on function search in module re:

search(pattern, string, flags=0)
    Scan through string looking for a match to the pattern, returning
    a match object, or None if no match was found.
```

This example shows how to conduct basic pattern search. The function `search` of the `re` module takes as input two strings: a pattern and a sequence and returns a match object. This is a special type defined in the module `re`. The match object represents the first occurrence of the pattern in the sequence. In this case, the pattern was the string `"hello"`. The method `group` returns the matching string, while the methods `start`, `end`, and `span` return its position within the sequence, as can be seen from the example.

Let's replace the pattern `"hello"` by `"[hH]ello"`. This is the way to say we are looking for either `"hello"` or `"Hello"`:

```
>>> match = re.search("[Hh]ello", "Hello world, hello Python!")
>>> match.span()
(0, 5)  <= reports only first match
```

Only the first match is reported. To find *all* the matches, not just the first one, we use a different method, termed `finditer`. It returns an object of type iterator, which is an object one can iterate over, using a simple "for" loop. The elements of this iterator are match objects, so we can apply the methods `group`, `span`, etc. to them.

```
>>> matches_lst = re.finditer("[Hh]ello","Hello dogs, hello cats!")
>>> for m in matches_lst:
        print(m.group(), m.span())

Hello (0, 5)
hello (12, 17)
```

Another, simpler way to get all the matches would be to call `findall`:

```
>>> matches_lst = re.findall("[Hh]ello","Hello world, hello Python!")
>>> matches_lst
['Hello', 'hello']
```

The `findall` method returns a list with all the matching strings. Since it does not return match objects, you cannot use `group` or `span` (in particular, it does not retain information on the location of the matches within the text string).

Exercise 1

Write a function `find_motifs` that receives a pattern and a sequence. The function will print all the matches, together with their starting and end positions within the sequence. At the end, the function will print the number of matches that were found.
   For example:

```
>>> find_motifs("[Hh]ello", "Hello dogs, hello cats!")
Hello from 0 to 5
hello from 12 to 17
2 matches found
```

4.2.2 Meta-Characters

Let us delve deeper into the syntax of Python's regular expressions. We already saw that by using `[]` we can define a pattern with several possibilities at a specific location. The

Table 4.1. *Meta-characters, meaning, and examples*

Meta-character	Meaning	Examples and comments
[ ] (matching square brackets)	a set of possible characters	"[ABC]" will match "A" or "B" or "C". "[A-Z]" will match any uppercase letter. "[a-z0-9]" will match any lowercase letter or digit.
[^...]	the complement of a set	"[^R]" will match any character except "R". re.findall("[^H]ello","Hello world, hello Python!") will match "'hello".
. (dot)	any single character, except new-line	"ATT.T" will match "ATTCT", "ATTGT" but not "ATTTCT".
| (vertical bar)	"or"	"ATT|CG|C" will match "ATT", "CG" or "C".
( )	used for grouping	"AT(T|CG|C)" will match "ATT", "ATCG" or "ATC".
*	0 or more occurrences	"AT*" will match "A", "AT", "ATT", but not "TT".
+	1 or more occurrences	"AT+" will match "GAT", "GATT", but not "GA".
?	exactly 0 or 1 occurrences	"AT?" will match either "A" or "AT".
{n}	exactly n occurrences	"(ATTG){3}" will match "ATTGATTGATTG" but not "ATTGATTG".
{m,n}	between m to n occurrences	"(AT){3,5}" will match "ATATATATAT" but not "ATATTATAT". Omitting m is interpreted as a lower limit of 0, while omitting n results in an upper bound of infinity.

characters "[" and "]" are called **meta-characters**: they have special meanings when contained in a pattern. Table 4.1 contains some of Python's meta-characters:

```
[ ]  ^  .  |  $  *  +  ?  { }  \  ( )
```

Exercise 2

Write a function `has_tata` that checks whether a given DNA string contains a TATA-box-like pattern ("TATAA" followed by three nucleotides and ends with "TT").

For example:

```
>>> has_tata("CCCTATAAAAATTCCC")
True
>>> has_tata("CCCCCCCCCCCCCCCC")
False
```

Table 4.2 contains two additional meta-characters that enable us to find a pattern that appears at the beginning or at the end of a sequence.

For example, suppose we wanted to check if a given genomic sequence is an ORF (open reading frame, which starts with a start codon, and ends with a stop codon) in bacteria.

Table 4.2. *More meta-characters*

Meta-character	Meaning	Examples and comments
^	the beginning of the sequence	" ^AUG" will match "AUGAGC" but not "AAUGC". Recall that using it inside [ ] means "opposite".
$	the end of the sequence	"UAA$" will match "AGCUAA" but not "ACUAAG".

The following would *not* do the work, since it does not enforce the sequence to *start* and *end* with these codons (it actually checks if `seq` *contains* an ORF):

```
1 def bad_is_orf(seq):
2     return re.search("ATG[ACTG]*(TAG|TAA|TGA)", seq) != None
```

So any sequence that contains an ORF would match the pattern. Let's fix it:

```
1 def is_orf_almost(seq):
2     return re.search("^ATG[ACTG]*(TAG|TAA|TGA)$", seq) != None
```

But this is still not quite right. The ORF must contain a number of nucleotides that is a product of 3, as each codon contains exactly three nucleotides. This, too, is easy to fix:

```
1 def is_orf(seq):
2     return re.search("^ATG([ACTG]{3})*(TAG|TAA|TGA)$", seq) != None
```

An alternative way to achieve this would be:

```
1 def is_orf(seq):
2     return len(seq)%3 == 0 and \
3            re.search("^ATG[ACTG]*(TAG|TAA|TGA)$", seq) != None
```

There is still a problem with this code, since a stop codon may appear in the middle, as in "ATGTAATAA." The function `is_orf` will return True, but we may want to exclude such strings from being considered reading frames. We will deal with this in the next section.

Finally, suppose you want to replace every occurrence of your pattern with another string. You can use `re`'s function `sub` (substitution). For example:

```
1 def replace_tata(seq):
2     return re.sub("TATAA[ACGT]{3}TT", "-tata-", seq)
```

```
>>> replace_tata("GGTTCGGATGATGTATAAGGCTTGGTATAAAAATTGGCTTATA")
'GGTTCGGATGATG-tata-GG-tata-GGCTTATA'
```

4.2.3 Length Matters!

Suppose we want to find all positions of ORFs within a sequence. Let's try this:

```
1 def find_all_orfs_greedy(seq):
2     orfs = re.finditer("ATG([ACTG]{3})*(TAG|TAA|TGA)," seq)
3     for orf in orfs:
4         print(orf.start(), orf.end())
```

Look at these executions:

```
>>> find_all_orfs("CCCATGCCCTAACCC")
3 12   <= Good!
>>> find_all_orfs("CCCATGCCCCCCTAACCCCCCATGCCCCCCTAACCC")
3 33   <= Not so good...
```

In the first execution, everything seems fine. However, in the second one, there seems to be a problem. The sequence contains two reading frames:

```
"CCCATGCCCCCCTAACCCCCCATGCCCCCCTAACCC"
```

So why did we find only one? If you check the positions found, 3 is the beginning of the first ORF and 33 is the end of the second ORF. Python indeed found an ORF, but not the shortest ORF in that sequence.

Another example:

```
>>> re.findall("a[ac]+c","aaaacccaaaaaccccc")
['aaaacccaaaaaccccc']
```

Repetitions meta-characters, such as * and + are **greedy**: the matching engine will try to find the longest repeat. Only if later portions of the pattern don't match, will the matching engine roll back and try again with fewer repeats.

Python allows us to conduct a non-greedy search. To find the minimal possible match, use *? (instead of *), and +? (instead of +). This way, Python will match as little text as possible. Suppose we wanted to find all positions of ORFs within a sequence. Let's try this:

```
1 def find_all_orfs(seq):
2     orfs = re.finditer("ATG([ACTG]{3})*?(TAG|TAA|TGA)", seq)
3     for orf in orfs:
4         print(orf.start(), orf.end())
```

Look at these executions:

```
>>> find_all_orfs("CCCATGCCCTAACCC")
3 12   <= Good!
>>> find_all_orfs("CCCATGCCCCCCTAACCCCCCATGCCCCCCTAACCC")
3 15
21 33 <= Very Good!
>>> re.findall("a[ac]+?c", "aaaacccaaaaaccccc")
['aaaac', 'aaaaac']
```

Table 4.3. *Classes of characters in Python's* `re`

String	Class	Equivalent
\d	Decimal digit	[0–9]
\D	Non-digit	[^0–9]
\s	Any whitespace[1]	[\t\n\r\f\v]
\S	Non-whitespace	[^\t\n\r\f\v]
\w	Any alphanumeric	[a-zA-Z0–9_]
\W	Non alphanumeric	[^a-zA-Z0–9_]

4.2.4 Classes of Characters

Python's `re` syntax defines several classes of characters, which appear in Table 4.3.

These classes provide shortcuts to popular sets of characters. For example, we can check if a string represents a floating-point number (a representation of a rational number, e.g., 3.14 or 100.0): it should start with at least one decimal digit, then a dot, then again at least one decimal digit. Here is the way to do it:

```
>>> re.search("^\d+\.\d+$", "3.14").span()
(0, 3)
```

4.2.5 Compiling a Pattern for Reuse

Compiling a pattern basically means **preprocessing** it, storing it in the computer's memory in some clever manner, in order to speed up future searches of this pattern. Preprocessing can be viewed as an "off-line" computation, done once on part of a problem's input provided to us beforehand in order to speed up subsequent computations given additional "on-line" input. Using compilation is advisable when we look for the same pattern many times within different sequences. This speeds things up, as an efficient representation of the regular expression is generated and stored in the computer's memory only once, and then reused again and again. Such preprocessing has some similarities with the binary search approach we encountered earlier: in order to efficiently search items within a list, we can sort the list first (this is the preprocessing, and needs to be done only once, as long as the list remains unchanged). As a result, each subsequent search will cost less in terms of running time.

Compiled REs also have the methods `search`, `findall`, and `finditer` among other methods.

---

[1] A whitespace is a character that does not correspond to a visible symbol, but typically does occupy an area when printed (e.g., space, tab, new line, carriage return, and several other characters).

For example:

```
>>> pattern = re.compile("[Hh]ello") # pattern is a "compiled" re
>>> pattern.findall("Hello dogs, hello cats!")
['Hello', 'hello']
>>> pattern.findall("Hello world")
['Hello']
```

Note that you can use the variable `pattern` again and again.

4.3 Finite State Machines (FSM)

We conclude this chapter with an explanation of **finite state machines** (FSM), a very basic model in computer science, closely related to regular expressions. FSMs may provide intuition into how regular expressions work. We note that RE matching in Python, as well as in most common tools and languages, involves many additional details that are not presented here.

A finite state machine is an abstract mathematical model. Let's start with a simple example for a FSM (see Figure 4.1), which includes two nodes. These nodes, denoted q_0 and q_1 are called **states**. The machine is at one of its states at any given moment. One of these states, q_0, is marked "start," and termed the **initial state**. The machine also defines several **transition rules**, described by the edges. When the machine is at a specific state, it can "read" a character from a defined set of possible characters, and possibly move to another state. For example, suppose the machine, which starts at q_0, reads the character "A" as the first letter of its input. Then the machine will change its state to q_1, as represented by the directed edge labeled "A." If, however, the machine at q_0 reads "C," "T," or "G," it would stay at q_0, as represented by the self-loop on q_0 labeled by these characters. We can imagine that a machine starts at its initial state, reads a sequence of characters, one at a time, and changes states according to its transition rules.

At some point, the sequence of characters will terminate (assuming the input sequence is indeed finite). When this happens, the machine will be at one of its states. You may notice that q_1 is colored yellow and denoted by a double circle. This is our way of representing the fact that q_1 is a "final" or **"accepting" state**. The meaning of this is that if the machine "rests" at state q_1 when it has finished reading its input, we say that it has **accepted** the input. Otherwise, we say it has **rejected** it. More generally, some of the machine's states are defined as being accepting. A machine accepts the strings that "lead" to an accepting state. In other words, imagine "walking the string through the machine," and declare acceptance if and only if you end up in an accepting state. For example, if we supply this FSM with the input string "AAAC", it will start at q_0, read an "A" and move to q_1, then read 2 more "A"s and stay and q_1, and then read a "C" and go back to q_0. So this FSM does not accept "AAAC," but it does accept, e.g., "AAA" or "CA".

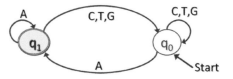

Figure 4.1. A simple example for a finite state machine

Table 4.4. *Transition rule table corresponding to the FSM in Figure 4.1*

from \ move to	q_0	q_1
q_0	C, T, G	A
q_1	C, T, G	A

The next step is to ask: What's common to all the strings that this FSM accepts? The answer is that they all must end with an "A." This is both a necessary and a sufficient condition – a string that does not end with an "A" will be rejected by this FSM. So, what we have here is a way to define patterns using a FSM, just like regular expressions. A fundamental result in computer science is that any pattern you can define by a regular expression can also be defined by a FSM, and vice versa.

Finally, it is worth mentioning that FSMs are also termed **Deterministic Finite Automata** (DFA), and an automaton[2] is another name for a state machine. "Deterministic" means that there is always a single state to move to, given the current state and the next input character. We note that there are also non-deterministic automata, but we will not explain these here.

Let us briefly summarize: a FSM is defined by:

- A finite alphabet Σ of input characters, for example $\Sigma = \{A,T,C,G\}$
- A finite set of states S, for example $S = \{q_0, q_1\}$
- An initial state, where the computation starts, for example q_0
- A set of one or more final ("accepting") states, for example $\{q_1\}$
- A set of transition rules (see example in Table 4.4).

Exercise 3

(a) Figure out what strings do the FSMs in Figure 4.2 accept (over $\{A,T,C,G\}$)?
(b) Write a corresponding regular expression.

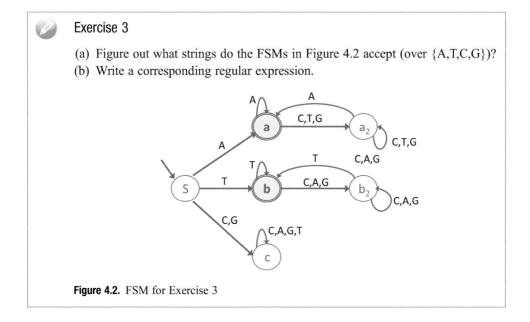

Figure 4.2. FSM for Exercise 3

---

[2] Automaton is the single form of automata. Just like other words of Greek origin, such as bacteria and bacterium, phenomena and phenomenon, data and datum.

Exercise 4

(a) Draw a FSM accepting all strings that contain a codon for Arginine. (CGT / CGA / CGC / CGG / AGA / AGG)

  Use as few states as you can (6 should suffice).

  Don't forget to specify the initial and accepting state(s).

(b) Give a regular expression in Python's syntax for this automaton.

FSMs are in fact another way to look at patterns, which are defined by regular expressions. They can also teach us something about the limitations of regular expression. For example, it turns out there is no regular expression for a pattern such as $a^n b^n$, that is, a specific (but unknown) number of 'a's followed by the exact same number of 'b's. The proof of this claim is omitted.

In addition to the importance of FSMs in theoretical computer science, these are excellent models used to describe various processes, such as controllers of simple devices. For example, a vending machine can be viewed as a FSM: each coin inserted moves it to another "state" in which the cumulative amount of money is remembered, until the user presses the button for a can of a desired drink. Then, the machine resets the amount to 0 (possibly after returning some change) and goes back to its initial state. Other examples for the modeling using FSMs include decision making, natural language processing, and of course pattern matching, as we saw.

Reflection

In this chapter, we introduced regular expressions (RE) – a way to represent textual patterns with possibly many "degrees of freedom," repetitions, etc. Python module re defines a "language" (or syntax) for the definition of regular expression, which is very similar to the syntax of other languages and existing tools. We did not cover all the details of this syntax. For a more thorough introduction, you may want to refer to Python's official tutorial. There are many tools for creating and working with regular expressions. One is Kodos (http://kodos.sourceforge.net), a GUI utility (written in Python), which complies with Python's re syntax.

Regular expressions do not convey any probabilistic information, e.g., on the probability of each possible alternative match. For example, the TATA-box regular expression represents any of the possible matches with equal probability, although in nature some are more common than others, and the likelihood of a specific TATA-box varies among species. Biologists sometimes use sequence logos (see Figure 4.3) to convey probability information for such strings. In sequence logos, the size of the character is correlated with its abundance at that specific position. Sequence logos are typically limited to very simple, fixed length patterns.

Challenge Yourself

Problem 1 Digestion by restriction enzymes

Suppose we have two restriction enzymes, called X and Y. The recognition site of X is ANT-AATTTTT, and of Y is GCRW-TGGGGG (N means any base, R means A, or G, W means A or T, and "-" indicates the position of the cut site).

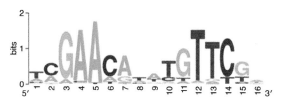

A sequence logo for the LexA-binding motif of several
Gram-positive species (from Wikipedia)

Figure 4.3. An example for a sequence logo

(a) Predict the fragment lengths that we will get, if we digest the genome of *Mycobacterium leprae* with X. For example, if the genome was "CCCAATAATTTTTGGGATTAATTTTTGG," then X would cut between the nucleotides underlined. We would get three fragments: "CCCAAT," "AATTTTTGGGATT," and "AATTTTTGG," whose lengths are 6, 13, and 9.

(b) Repeat the previous section when the two enzymes X and Y digest the DNA sequence together.

Problem 2 Printing the "context" of a pattern

(a) Write a function `find_pattern(seq, pattern, pre, post)`, which takes a sequence `seq`, a pattern `pattern` (a string in the syntax of Python's regular expressions), and two additional integers `pre` and `post`. The function will print all the occurrences of `pattern` in `seq`, in their local "context": the local context of a pattern begins `pre` positions before and ends `post` positions after the pattern (if the context outreaches either ends of `seq`, fewer characters will be printed). The pattern will be printed in uppercase letters, while the context in lower case letters. In addition, the beginning position of the pattern will be printed.

   For example: on the book's website we have a file cap2.txt, which contains the protein Adenylyl cyclase-associated 2 (CAP2) of the Orangutan in FASTA format:

```
>Q5R5X8|CAP2_PONPY CAP 2 - Pongo pygmaeus (Orangutan).
MANMQGLVERLERAVSRLESLSAESHRPPGNCGEVNGVIGGVAPSVEAFDKLMDSMVAEF
LKNSRILAGDVETHAEMVHSAFQAQRAFLLMASQYQQPHENDVAALLKPISEKIQEIQTF
RERNRGSNMFNHLSAVSESIPALGWIAVSPKPGPYVKEMNDAATFYTNRVLKDYKHSDLR
HVDWVKSYLNIWSELQAYIKEHHTTGLTWSKTGPVASTVSAFSVLSSGPGLPPPPPPPPPP
```

   After reading it to a variable CAP2 and eliminating "\n" characters, we get:

```
>>> find_pattern(CAP2, "NM", 5, 5)
maNMqglve starts at position 2
rnrgsNMfnhls starts at position 127
```

(b) The CAP2 protein contains the following pattern: the first two positions are amino acids among L, I, V, or M, then any amino acid, then R, then L, then one of D or E, the next four positions contain any amino acid, and at the end R, then L, and finally E.

   Find the occurrence of this pattern and its position within CAP2. Use your function from (a).

Part III: Graphs and Networks

– "Connecting the dots."

Graph theory is a central field in mathematics and computer science. The word "graph" has different meanings in different contexts. In high school, when you say "graph," you usually mean a curve that visually describes a function (e.g., $y = x^2$). In economics and statistics, graphs are diagrams that describe relationships and trends between variables. By way of contrast, in pure mathematics and computer science, the word "graph" has a completely different meaning.

A graph is defined by two sets. One set, the **nodes**, describes a collection of objects, for example biological species, such as human, whale, mountain lion, crow, pigeon, sea gull, Arabian surgeonfish, zebra fish, and Anemonefish. The second set, the **edges**, comprises pairs of nodes, such as {human, mountain lion}, {pigeon, sea gull}, {Arabian surgeonfish, Anemonefish}. Each pair represents a connection between objects, which has some meaning. For example, two species are related to one another if they belong to the same taxonomic family, or they both live on the same continent, or they are active in the same part of the day, and so on. Each such feature will create a different graph, because different properties create different sets of pairs (edges).

One can draw a graph as a set of points or circles connected by lines. For example, the drawing in Figure III.1 illustrates a particular graph, known as the Paterson graph, which has 10 nodes and 15 edges.

Clearly, it is feasible to draw only graphs with a fairly small number of nodes and edges. Consider, for example, the Facebook graph, whose nodes are people with a Facebook account, and there is an edge between two people if they are Facebook friends. This graph has over two billion nodes, and obviously, there is no way to draw it graphically. Statistics and properties of this graph, such as nodes with a large number of incident edges, and distances between members in specific groups are of much interest to Facebook itself, as well as to researchers interested in the sociology and dynamics of the internet, companies interested in targeted advertisements, political candidates, and many others.

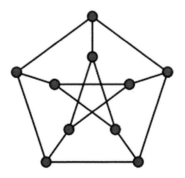

Figure III.1. A Paterson graph

Graphs are very common in computational biology as a means to represent various types of biological interaction networks, such as protein–protein networks (PPI), transcription regulation networks, and metabolic networks.

In this book, we aim to expose the reader to some basic notions in graph theory (chapter 5), including directed and undirected graphs, paths and connectivity, and clustering. In chapter 6, we touch upon algorithms that are applied on graphs, particularly breadth-first search (BFS), and in chapter 7 we introduce the notion of simulation and exemplify how graphs can be used to simulate biological regulatory networks.

5   Basic Notions in Graph Theory

This chapter begins with a definition of what a graph is. While the definition is rather abstract, it is general enough to make graph theory relevant and applicable to a wide range of systems, in a variety of contexts. We will also see how to represent graphs in Python (or any other programming language). Such a representation allows a computerized inspection of many graph properties, as well as implementing various algorithms on graphs. By the end of this chapter, you will be familiar with many basic notions in this field. The last two sections of this chapter introduce the notions of clusters and clustering, and of hierarchical clustering in the context of phylogenetic trees. We note that the treatment of these two topics here hardly touches their surface – both are deep subjects in their own right, covered by vast and diverse literature.

5.1 Graphs: Definitions

The basic elements in a **graph** are called **nodes**. A graph consists of a set of nodes and a set of **edges**, where an edge connects two nodes. The set of edges of a graph can be empty, full (i.e., contain all possible edges), or be any subset in between. Nodes can represent many types of entities, making graphs a highly diverse model for many phenomena. For example, the graph in Figure 5.1 consists of seven nodes, which represent protein coding genes, involved in the yeast meiosis. Edges in that graph correspond to interactions that occur between the resulting proteins. Note that there are no directions to the interactions here. Therefore, the edges are pictured as undirected lines, and not as directed arrows. This graph is thus **an undirected graph**.

Formally, a graph G consists of two sets V and E. V is a set of nodes (also called **vertices**), and E is a set of edges. The set E itself consists of pairs of nodes from V. In Figure 5.1,

$V = \{$DBF20, UME6, IME1, IME2, RIM11, STH1, HOP2$\}$, and
$E = \{$(DBF20, UME6), (DBF20, IME1), (UME6, RIM11), (UME6, IME1), (UME6, IME2), (IME1, IME2), (IME1, RIM11), (IME1, STH1), (IME2, STH1)$\}$

Note that this is a mathematical definition, having nothing to do with Python sets or tuples. We say that an edge "connects" the two nodes adjacent to it, or that it "touches" them. The number of nodes is usually denoted n (here $n = 7$) and the number of edges m (here $m = 9$).

When there is an edge connecting some nodes v_1 and v_2 we say that v_1 and v_2 are connected, adjacent, or are **neighbors**. The **neighborhood** of a node is the set containing all its neighbors. For example, the neighborhood of node IME2 in Figure 5.1 consists of IME1, STH1, and UME6. The neighborhood of node HOP2 is empty. The **degree** of a node v, denoted $deg\,(v)$, is the number of neighbors it has. For example, deg(IME2) $= 3$, and deg(HOP2) $= 0$.

The relations represented in a graph can be directional. In this case, node v_1 may affect node v_2 in some way, but not vice versa. Such graphs are termed **directed graphs** (or digraphs). For example, regulatory networks are represented as directed graphs, in which a

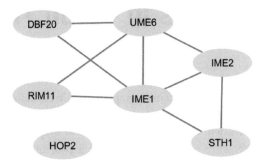

Figure 5.1. An undirected graph, representing protein interaction in yeast meiosis

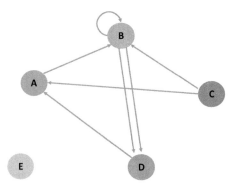

Figure 5.2. A directed graph

node activates or inhibits other nodes. In directed graphs, when we have an edge from node v_1 to node v_2, we call v_1 the **source** and v_2 the **target** of the edge. Also, in directed graphs we separate the **in-degree** and **out-degree** of a node. These are the number of incoming and outgoing edges to and from the node, correspondingly.

Sometimes a source node may be connected to a target node with two or more edges. Such edges are called **parallel edges** (see the edges from B to D in Figure 5.2). When the source and target node are identical, we call the edge a **self-loop**. For example, node B in Figure 5.2 has a self-loop. Self-loops can be used to model, e.g., proteins that activate or inhibit their own production or activity. The term **simple graph** relates to graphs with neither parallel edges nor self-loops.

5.2 Weighted Graphs

It is often desirable to assign **weights** to edges of graphs. This is applicable both to undirected and directed graphs. Such graphs are called **weighted graphs**. The weights that are assigned to edges are typically real numbers. Generally, weights can represent various types of information, such as lengths of roads, capacity of a cable connection, strengths of chemical bonds, resistance of an electrical wire, and capacity of a channel. In a biological

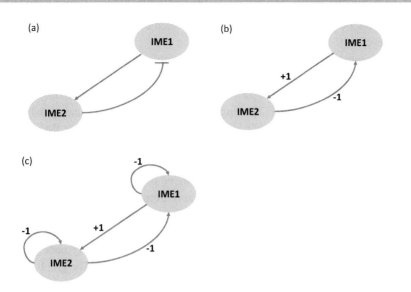

Figure 5.3. (a) Regulation network of proteins IME1, IME2 (b) Weighted graph equivalent (c) Two self-loops added, depicting self-inhibition

context, particularly in regulation networks, positive weights stand for activation effects, while negative ones stand for inhibition. For example, Figure 5.3(a) represents a simple regulation network, in which the protein IME1 activates IME2, and IME2 inhibits IME1. Figure 5.3(b) shows how to represent this as a weighted graph: the edge from IME1 to IME2 is assigned weight +1, and the edge in the opposite direction is of weight −1. If, in addition, we know that IME1 and IME2 degrade over time even without any external effect by other proteins, we can represent this by adding self-loops with negative weights on both nodes, as shown in Figure 5.3(c).

5.3 Paths and Connectivity

A **simple path** in an undirected graph is a sequence of connected, distinct nodes (so each node appears just once). For example, in Figure 5.1, there is a path from DFB20 to RIM11, which goes through IME1. We denote this path $p = $ (DFB20, IME1, RIM11). In the context of a protein interaction network, such as the one in Figure 5.1, the existence of a path between two nodes often has a biological meaning. It may represent some indirect interaction between the nodes along the path (when the path consists of just a single edge, the interaction is direct). A **cycle** is similar to a path, only it starts and ends at the same node. For example, cycle $p = $ (DFB20, UME6, IME1, DFB20) in Figure 5.1.

   We can similarly define paths and cycles in directed graphs. In a directed path, we can move only along the valid direction across each edge. For example, consider the directed graph in Figure 5.2(a) directed path from A to D is $p = $ (A, B, D) but (A, D) is not a path from A to D because the direction of the edge is from D to A. Another example of a cycle is

$p = (A, B, D, A)$, but (A, B, C, A) is not a valid directed cycle. A graph is called **acyclic** if it contains no cycles.

The length of a path p, denoted $|p|$, is the number of edges in it. In weighted graphs, the weight of a path is the sum of weights of edges along the path, denoted $w(p)$. Sometimes we use $w(p)$ to denote the length of a path in non-weighted graphs as well. This complies with setting the weight of all edges to 1. The length and weight are defined analogously for cycles. For example, in Figure 5.1, the length of the cycle $p = $ (DFB20, UME6, IME1, DFB20) is $w(p) = 3$.

In unweighted graphs, the **distance** between two nodes v_1 and v_2, denoted $dist(v_1, v_2)$ is the minimal length of a path from v_1 to v_2. For example, in Figure 5.1, $dist$(DBF20, RIM11) $= 2$, since RIM11 is reachable from DBF20 using two edges, and no shorter path exists. If v_2 is not reachable from v_1, then the distance is defined as infinity, e.g., $dist$(HOP2, IME1) $= \infty$.

Finally, a graph is called **connected** if there is a path from every node to every other node (in other, words, every node is reachable from any other node). The graph in Figure 5.1 is not connected, but removing the node HOP2 from it will turn it into a connected graph. For directed graphs, we require a directed path between every two distinct nodes.

In the next chapter, we will see an efficient algorithm that computes the distances between nodes in a non-weighted graph. This algorithm can be used to find out whether a given graph is connected or not.

5.4 Eulerian and Hamiltonian Paths

You may be familiar with the following riddle, called *"The House of Santa Claus"*: is it possible to draw the house depicted in on the left of Figure 5.4 in one stroke, without raising the pen from the paper, and without redrawing any edge?

This type of problem can be analyzed using tools from graph theory. Figure 5.4 on the right shows a graph that models that house (possibly omitting some artistic features). What we ask is whether there exists a path in the graph that goes through every edge exactly once. Such a path is known as a **Eulerian path**, named after Leonhard Euler (1707–1783), the Swiss mathematician, physicist, astronomer, geographer, logician, and engineer. It may take a while to find such path, but it does exist. For example, see the graph in Figure 5.4, where the numbers on the edges indicate the order of edges in such a path (the lower left node is the starting point of the path, and the lower right is its end).

Euler was actually interested in a related problem, known as *"The seven bridges of Königsberg."* In the city of Königsberg, today's Kaliningrad in Russia, there is a river that

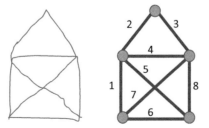

Figure 5.4. The house of Santa Claus and its corresponding graph

(a)   (b)

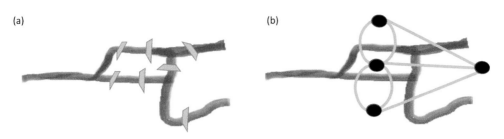

Figure 5.5. The seven bridges of Königsberg
(a) The bridges split the town into four parts – northern, southern, middle, and eastern
(b) The corresponding graph

used to split the town into four parts, and getting from one part to another requires crossing a bridge (see Figure 5.5(a)). Euler tried to figure out if it was possible to cross all the bridges in the city, without crossing any bridge more than once. This layout of the bridges can be modeled as an undirected graph with parallel edges, as shown in Figure 5.5(b).

Euler characterized the conditions for a graph to contain a path or a cycle that goes through every edge exactly once. These conditions are sufficient and necessary: when satisfied, the graph will contain a Eulerian path/cycle, and vice versa – any graph that contains such a path/cycle must satisfy the condition. This characterization is both simple and elegant:

Theorem (Euler, 1736)

(a) A connected undirected graph (possibly with parallel edges) contains a Eulerian cycle if and only if the degree of every node is even.
(b) A connected undirected graph (possibly with parallel edges) contains a Eulerian path if exactly two nodes have an odd degree. Furthermore, the path must begin in one of them and end in the other one.

We will not formally prove this theorem, but give some intuition instead. A Eulerian cycle starts with any node and returns to it in the end. Along the cycle, we "step in" and "step out" of nodes. Each such step uses a single edge in the graph. Note that we must "step in" and then "step out" of each node the same number of times. This is true for every node in the graph, including the starting one (because it is also the end node). Therefore, the number of edges that touch each node must be even.

The generalization to a Eulerian path is similar, only the number of times the path goes in and out of first and last nodes is not the same as in a cycle. The path leaves the first node one more time than entering it, and for the last node, the path enters it one more time than leaving it. Thus, the degree of these two nodes is odd.

Using this characterization, Euler proved that it is not possible to go through the seven bridges of Königsberg in the defined manner. The corresponding graph has three nodes of degree 3 and one node of degree 5. Therefore, the conditions for a Eulerian path, or cycle, do not hold.

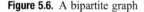

Exercise 1

(a) Explain why the graph in Figure 5.4 contains a Eulerian path, but not a Eulerian cycle.

(b) Add one edge to the graph in Figure 5.4, such that it will contain a Eulerian cycle.

A closely related problem asks if a given graph has a path (or cycle) that visits every *node* exactly once. Such a path is called a **Hamiltonian path**, after physicist William Rowan Hamilton (1805–1865). As we will see in a later chapter in the book (Mission Infeasible), this problem is surprisingly difficult to solve, despite its similarity to the Eulerian paths problem. Same, same, but different!

5.5 Special Types of Graphs

We briefly describe three special types of graph, which are often used in biological research.

The first type is a **bipartite** graph. This is a graph whose nodes can be partitioned into two disjoint sets, such that all the edges "cross" between these two sets (see Figure 5.6). Graphs representing enzyme–substrate relations, gene–disease relations, and transcription regulation networks are often bipartite. In the latter, for example, there are two types of nodes – nodes representing genes that encode transcription factors, and nodes representing their products – the transcription factors. Edges in such graphs typically connect only nodes of the two different types.

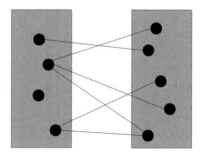

Figure 5.6. A bipartite graph

Exercise 2

(a) Is the graph in Figure 5.7 bipartite?

(b) (a more difficult challenge) Can you specify a condition, which is both necessary and sufficient, for a graph to be bipartite?

Hint: look at which cycles are included in the graph.

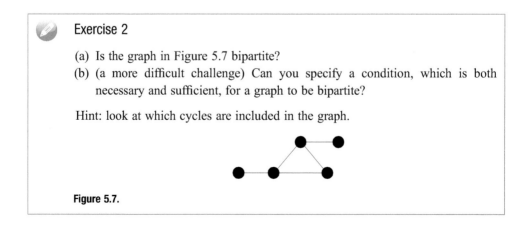

Figure 5.7.

The second special type of graph is a **clique**, or a **fully connected graph**. This is a graph containing all the edges between all pairs of nodes (see Figure 5.8), usually not including self-loops. Cliques in biological networks are the ultimate, most dense case of highly connected subnetworks, often having a specific, designated functionality.

Exercise 3

How many edges are there in a clique of size n (the number of nodes),

(a) Not including self-loops?
(b) Including self-loops?

The third special type of graph, depicted in Figure 5.9(a), is a **tree**. Here, a tree will refer to undirected trees, unless specified otherwise. A tree is a connected graph with the following properties: 1) removing any edge from it will turn it into a non-connected graph, and 2) adding any edge to it will introduce a cycle. In other words, a tree is a graph that is both connected and acyclic (contains no cycles). The tree in Figure 5.9(b) is a **rooted tree**: a tree with a special designated node, called the **root**, which induces a hierarchical structure. If the number of nodes in the tree is two or more, then the root has "children", which may have children of their own, etc. Nodes without children are termed **leaves**. Curiously enough, computer scientists draw rooted trees such that the root is at the top and the leaves are at the bottom. Switching our attention to unrooted trees, they are still connected and acyclic, but they do not have a root, and thus no unique orientation, or hierarchical structure.

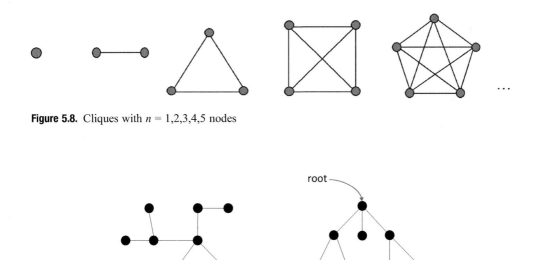

Figure 5.8. Cliques with $n = 1,2,3,4,5$ nodes

Figure 5.9. (a) An unrooted tree (b) A rooted tree

5.6 Representing Graphs

There are two common and simple ways to represent graphs in the computer's memory: adjacency matrices and adjacency lists. We will demonstrate the two representations in Figure 5.10 that shows an undirected graph, whose nodes are numbered 0, 1, 2, 3.

For a simple graph with n nodes, an **adjacency matrix** is an n by n matrix. Entry i, j in the matrix (row i and column j) is 1 if there is an edge between node i and node j, otherwise it is 0. Thus, the adjacency matrix corresponding to the graph G in Figure 5.10 is

```
G = [   [0,1,1,1],
        [1,0,1,0],
        [1,1,0,0],
        [1,0,0,0]   ]
```

For example, $G[1][2] == 1$, since there is an edge between node 1 and node 2, while $G[3][1] == 0$ since node 3 is not connected to node 1. Since the graph in this example is undirected, $G[i][j]$ must be equal to $G[j][i]$ (why?). We call such a matrix *symmetric*. In directed graphs, it is certainly possible that entry i, j will be different from entry j, i.

If you are worried that such a representation will work only when the graph nodes are labeled with integers 0,1,2, and so on – don't be! We can simply assign an arbitrary enumeration to the graph nodes.

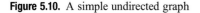   **Exercise 4**

In directed graphs, what does the i-th row represent? And the j-th column?

The other common graph representation method is an **adjacency list**. In Python style notation, we have a list whose elements are lists themselves, where the i-th inner list contains the neighbors of node i:

```
G = [   [1,3,2],
        [2,0],
        [0,1],
        [0]       ]
```

For example, $G[2]$ contains the neighbors of node 2 (nodes 0 and 1).

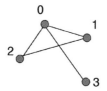

Figure 5.10. A simple undirected graph

Exercise 5

What do you think are the advantages of each representation over the other? When would you prefer each? Consider, for example, the following aspects:

(a) How would you include weights on edges in each representation?
(b) Which representation is more efficient memory-wise?
(c) Which representation is more efficient for checking if two given nodes are connected?
(d) Which representation is more efficient for printing all neighbors of a given node?

5.7 Random Graphs

Random graphs are graphs where edges, and sometime nodes, are generated by a probabilistic process. The study of random graphs in mathematics is an active topic, and many interesting, deep results were discovered and proved. Random graphs turned out to be a good way to model various physical, chemical, biological, and social phenomena, e.g., phase transition from ice to water, connections of neurons in the brain, vehicle traffic, and patterns of acquaintances among people. Many studies compare experimentally deduced networks to random networks, obtained by appropriate probabilistic processes. Such comparisons provide insights regarding the properties of real networks. For example, when a specific structure (e.g., a clique of a certain size) appears in a biological network much more often than it would appear in a random graph, this hints at some biological role, which may have been preserved through evolution.

There are various models for random graphs, and we discuss just two of the most common ones here. Perhaps the simplest model for random graphs is the **Erdős–Rényi** random graph model, introduced in 1959. We start with an empty undirected graph with n nodes, and connect every pair of nodes with probability p, independently. Two extreme cases are $p = 0$, where the resulting graph has no edges, and $p = 1$, where we get a complete graph (a clique on n nodes). The properties of such graphs as a function of p has been studied extensively by mathematicians, with some interesting results. Typical questions in such random graphs, given n and p, are: What is the size of the largest connected sub-graph? What is the expected size of the largest clique? Does the graph contain a Hamiltonian cycle? Many questions of this type are well understood by now, yet there are still interesting open problems.

Exercise 6

Complete the following function `rand_graph(n,p)` that takes the number of nodes n and the probability for each edge p, and returns a random undirected graph, in a matrix representation. Recall that random.random() returns a (pseudo) random floating point number between 0 and 1.

```
import random
def rand_graph(n,p):
    G = [[0]*n for i in range(n)]  # an n by n matrix of 0's
    for i in range(n):
        for j in range(i,n):
            r = random.random()
            if_____:

            _____

    return G
```

In Figure 5.11(a) below, you can see the graphical results for running the code above for generating a random graph with $n = 10$ and increasing values of p. Higher values of p typically yield a larger number of edges. Note that this holds with high probability, but not with certainty. In Figure 5.11(b), you can see that repeating the graph construction with the same n and p (this case, $n = 10$, $p = 0.4$) almost surely generates a different graph in each run.

Another type of a random graph model is called the **preferential attachment**, also known as the "rich-get-richer" model. Such a model was defined by Yule in 1925 in the study of the emergence of new species. The model defines a random process that generates graphs, in which the nodes with higher degrees are more likely to be further connected to additional nodes. The model was later studied by Barabási and Albert (1999). It is claimed that such random graphs better resemble natural and human-made networks, such as biological networks and the internet, compared to the Erdős–Rényi model.

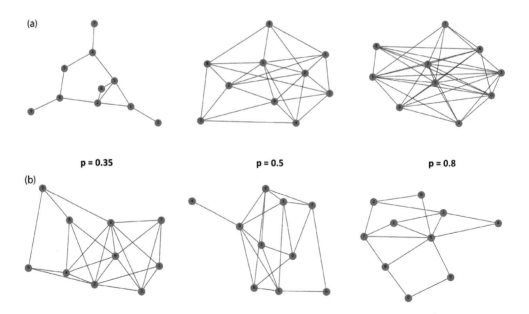

Figure 5.11. (a) Generating random graphs with $n = 10$ and increasing values of p
(b) Generating random graph with $n = 10$, $p = 0.4$ three times

5.8 Clusters and Clustering

Suppose we are given a set of objects, each characterized by a number of properties. We are often interested in partitioning these objects into different sets, such that objects within the same set are similar, or have similar properties, while objects from different sets are more distinct. Such sets are often called **clusters**, and the partitioning process is called **clustering**. When the clusters are pre-defined, the process is actually a classification of the nodes of a graph into those predefined clusters. When the clusters need to be deduced from the entire data, this is called unsupervised clustering (yet, we do have some information on the relations between the objects, on which the clustering is based: for example, we know which pairs of objects are related, or the pairwise distances between objects).

Clusters and clustering are important in many fields and contexts. In molecular biology, we often cluster genes according to their expression levels in different settings and environments. One important application of such clustering is the so-called "guilt by association" principle: we cluster genes by similar expression patterns in certain environments. Furthermore, we know the functionality of some genes in a cluster, but not the functionality of all of them. Clustering may help us get some initial idea about the unknown functionality of such genes, based on the functionalities of the other genes in the same cluster. In social sciences, clusters are often called communities, and the identification of such communities is an important building block in these fields. With the advent of online communities, such as Facebook, Twitter, or Amazon, the underlying graph becomes huge, containing billions of nodes (users/accounts). The detection of communities enables, for example, targeted and effective advertisements and dissemination of relevant information.

There are numerous clustering algorithms. Many of them rely on the ability to embed the objects in a metric space, where the distance between pairs of objects is given. Other families of clustering algorithms supply, for each pair of objects, the information on whether they are related or not. This induces an undirected graph, where edges represent pairwise similar objects.

A natural approach to clustering is to generate clusters by partitioning the nodes of the graph into cliques. Nodes in the same clique are all related to one another, and each node in the graph belongs to exactly one clique. Such partition is known as **clique cover**.

Figure 5.12 depicts a graph with seven nodes, which are covered by two cliques, one of size 3 (nodes A, B, and C, and green edges) and one of size 4 (nodes D, E, F, and G, and purple edges). Note that we cover all the nodes, but not necessarily all the edges.

We note that typically a graph may have more than one possible clique cover, and some of those may be non-informative and not useful. For example, if every clique is a single node (called a singleton), then although this is a legal a clique cover, it is neither interesting nor useful.

Exercise 7

Find alternative clique covers for the graph in Figure 5.12. Is there another clique cover with only two cliques? Is there a cover with three cliques? Is there a cover that uses more edges than the one depicted in the figure?

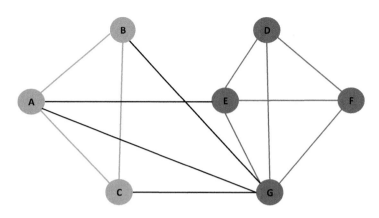

Figure 5.12. Clique cover of size 2

It is natural to apply some optimization criteria in order to select among possible clique covers. Optimization criteria refer to some measure, which we should either maximize or minimize. For clique cover, one such criterion is minimizing the number of cliques in the cover. We remark that finding a clique cover with a minimum number of cliques is a computationally hard problem. The fastest known algorithm that solves it runs in exponential time (see more on this in the "Mission Infeasible" chapter).

When algorithms that solve the problem optimally are not known, we often compromise on **heuristics**: algorithms that are not guaranteed to produce perfect (optimal) results, but in many cases of interest produce results that are good enough. In the clique cover problem, we can use the following heuristic:

Clique cover heuristic

1. Start: each node is a separate cluster (of size 1)
2. Repeat the following, until no two cliques can be merged:
3. For each cluster C in the current set of clusters:
    3.1 Find other clusters that are fully connected to C (all node in C are connected to all nodes in the other cluster)
    3.2 Among these, identify the largest other cluster, denoted D
    3.3 Merge C with D creating a new, larger cluster

We started with a collection of n singleton cliques, each containing a single node of the graph, and then "fuse" clusters as described. As this is a heuristic, the resulting clusters are often not optimal, with respect to the criteria defined above (minimal number of cliques that cover the graph nodes).

The heuristic we presented here is **greedy**. A greedy (or myopic) algorithm is typically iterative, and in each iteration it progresses based on the local optimal choice, namely the one which has the best results at the current configuration. Greedy algorithms ignore the "bigger picture," possibly missing out better solutions. However, they are usually simple and computationally efficient. There are actually cases where greedy algorithms do end up in optimal solutions. More often, the result is not optimal but is the best way we know to solve the problem, albeit not optimally.

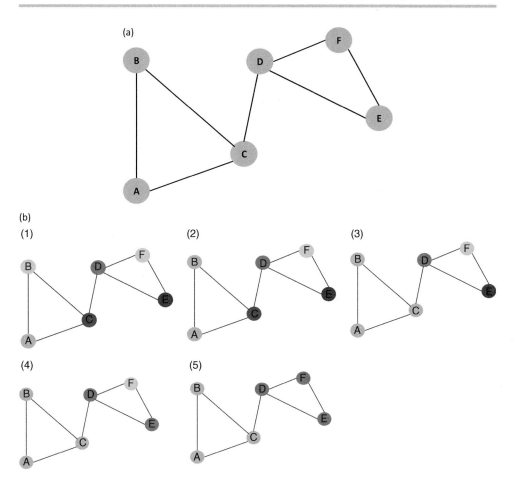

Figure 5.13. (a) An example graph for the illustration of the clique cover heuristic
(b) A possible algorithm run, resulting in two cliques, each of size three. Nodes' colors represent clusters.

Consider the graph in Figure 5.13(a). We show that running the above heuristic in different orders may yield different outcomes. For a start, examine a possible run in Figure 5.13(b), which terminates after 5 iterations.

Now to examine the second run in Figure 5.14. By the end of this run, there are three cliques of size two. You can see that changing the order not only affected the composition of the cliques, it also changed their number! With this order, the greedy algorithm failed its mission of giving the best clique-cover.

In a broader context, we note that often requiring clusters to be "perfect" cliques may be "too much" to ask for. Dense clusters (having many edges, but possibly not all edges) can be meaningful enough. In this context, a reasonable criterion for clustering is the so-called *maximizing agreements*: maximizing the number of edges within the clusters, while minimizing the number of edges between them. Intuitively, this calls for "dense" clusters. Unfortunately, the maximizing agreement problem is also computationally hard.

We conclude this section by a brief description of another very popular and important clustering heuristic, called **k-means**. The underlying context is of n objects, embedded in a metric space, so that each point is specified by a vector of coordinates.

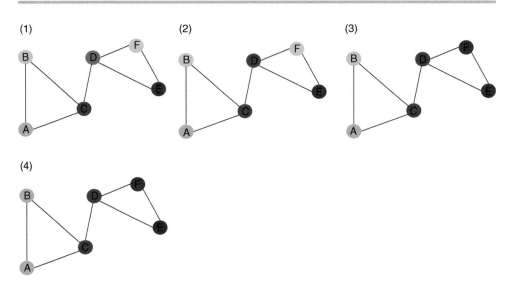

Figure 5.14. Another possible algorithm run, resulting in three cliques, each of size two.

We are given the number k, which is the number of clusters to be produced (typically, k is much smaller than n). Initially, we choose at random k distinct points out of the n points. These points are called *centers*. Then, for each point of the n points, we find the center that is closest to it (break ties at random). We then associate with each center the set of points closest to it. These k sets of points are the initial clusters.

We now iterate the following: we compute the "center of mass" of the points in the set. Intuitively this is the "average point" in the cluster. This average point is not necessarily any of the original points in the cluster. Now we have k updated centers. We associate each of the n points with their closest center, thereby forming new clusters, etc.

This iterative process is continued till some stopping condition is met. For example, such a criterion could be stabilization: no change in the composition of the clusters between two successive iterations. Note that due to the initial random choice of centers, this process is probabilistic. Different executions may well generate different clusters. In addition, in many cases it is not clear how k, the number of clusters, should be chosen in the first place. Indeed, the quality of the resulting clusters may well depend on k. In many cases, one simply searches over a range of possible values of k. With these reservations in mind, we note that k-means is a very efficient heuristic. In cases where there is a well-defined underlying structure of clusters, k-means is likely to find it. If the initial points cannot be partitioned into clear clusters, no miracles will occur, and the result will be of low quality. To summarize the algorithm:

k-means heuristic

1. Start: randomly choose k centers (k points from the n given points we wish to cluster)
2. Repeat the following, until there is no change in the composition of the clusters:
   2.1 for each point, define its cluster as the center closest to it
   2.2 for each cluster, recalculate a new center that is "in the middle" between its members.

Figure 5.15 depicts the dynamics of k-means clustering for objects (points) in the two-dimensional plane. Points are circles, centers are rectangles, and each cluster has a different color. Even numbered iterations show how centers are updated, while odd numbered

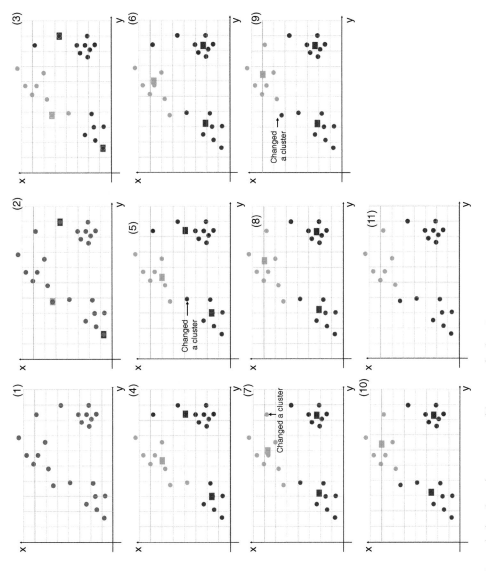

Figure 5.15. Example for clustering process of k-means, for $k = 3$

iterations depict the re-assignments of points to the updated centers. Since there is no change between the 9th and 10th iteration, the outcome is determined. We note that k-means clustering is frequently used in computational biology. For example, in clustering genes by expression levels, proteins by their activity, etc.

5.9 Hierarchical Clustering and Phylogenetic Trees

The type of clusters considered in the previous section are non-hierarchical. Every item belongs to a single cluster. In many contexts, it makes sense to consider **hierarchical clustering**. Here, an object belongs to more than a single cluster. Clusters are hierarchical with respect to set containment. An important property of such clusters is that, for any two clusters A and B, they are disjoint: A is a subset of B, or B is a subset of A. The hierarchy is determined by the containment relations among subsets. This naturally induces a rooted tree of subsets, where the root of the tree is the set of all objects under discussion.

Figure 5.16 illustrates non-hierarchical and hierarchical clustering. There are 6 points in a 2-dimensional space. We use Euclidian distance as a distance metric in both.

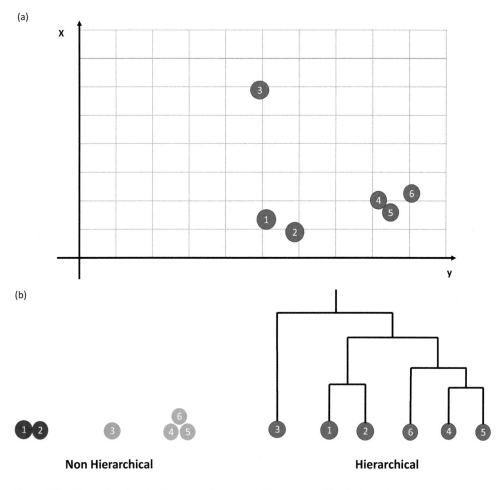

Figure 5.16. (a) 6 points in a 2-dimensional space (b) The results of both clusterings

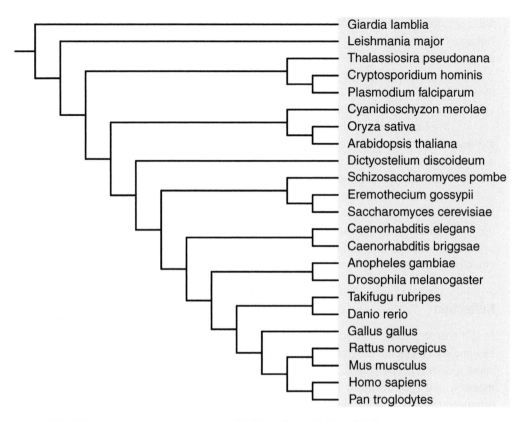

Figure 5.17. Eukaryota tree, as was generated with iTOL (https://itol.embl.de/)

Hierarchical clustering and trees are frequently used for reconstructing **phylogenetic trees**. Phylogenetic trees represent the evolutionary relationships between extant and extinct species. In the past, the input for the construction was based exclusively on fossil records. These days, molecular data (aligned sequences of amino acids or DNA nucleotides) are primarily used.

Most known species are classified both according to broader terms (such as Insecta) and narrower ones (such as Primates). Each term can be viewed as a cluster, containing a subset of all species. Any two distinct clusters A and B have the following property: they are either disjoint (empty intersection), (e.g., insects and primates), or one of them is a subset of the other (e.g., mammals, a subset of vertebrates). This directly corresponds to a rooted tree, with the set of all species represented by the root of the tree. Each subtree corresponds to a different cluster. For example, at the level just below the root, we have three clusters – Bacteria, Archaea, and Eukarya. These clusters are refined as we go deeper down the tree. At the very bottom, a leaf in this tree is a cluster of size one, corresponding to a single species (see Figure 5.17 for an example of a real hierarchical tree depicting the Eukaryote domain).

Consider the tree of life. The objects to be clustered are species. For example, Homo Sapiens (human), Drosophila melanogaster (fruit fly), or Yersinia pestis (a bacteria). When considering some specific (and rather arbitrary) species, Homo Sapiens belong to the Primates cluster (along with, say, chimpanzees), to the Mammalia cluster (along with, say, cows), to the Vertebrata cluster (along with, say, zebra fish), and to the Eukarya cluster (along with, say, fruit flies). Drosophila belong to Insecta (along with, say, wasps and moths), and Eukarya (along with humans). We finally turn to the Yersinia pestis bacterium (which causes the plague). It belongs to the Enterobacterale family, which is a subset of Proteobacteria, which in turn are a major phylum of gram-negative bacteria. Note that in describing these three paths of clusters (for human, fruit fly, and the plague bacterium), we have carried out a very severe reduction of additional, taxonomical clusters along the three paths.

We have briefly described the notion of hierarchical clustering, and its relevance to the tree of life. There are many algorithms and heuristics for phylogenetic reconstruction. These include distance-based methods (assuming pairwise distances among extant species are given, or can be reliably computed); sequence-based methods, such as maximum parsimony and maximum likelihood, super tree methods, Bayesian methods, and many others. Most of these methods attempt to maximize or minimize a certain objective function. In most cases, this is a hard computational problem, yet often the heuristics seem to produce rather satisfactory outcomes, even if not necessarily optimal. There is a vast literature on phylogenetic reconstruction methods, so we chose not to elaborate on any of them in this text.

Reflection

In this chapter, we introduced basic notions in the important field of graph theory: directed and undirected graphs, weighted graphs, degree of nodes, paths and connectivity, and probed some special types of graphs. We showed how graphs can be represented in the computer's memory, and specifically in Python, e.g., as a matrix and as an adjacency list. We also discussed random graph models, and touched upon the important topic of clustering.

Graphs are extremely useful in modeling various phenomena in a variety of areas. An example of some interest is the Facebook graph, where nodes are users, and there is an undirected edge between two users if they are Facebook friends (not to be confused with real friends, of course). Notions like the distance between specific users, the average distance between users, and hubs of high degree (see exercise, below), are of interest when trying to understand the social structure underlying this graph. Similarly, there is an interest in partitioning the graph to clusters (or to hierarchical clusters) and correlating them to other attributes such as area of residence, profession, and age group.

After becoming familiar with these notions, we are ready to move a step further. In the next chapter, we will provide a glimpse into *graph algorithms* – the study of algorithms whose input is a graph.

Challenge Yourself

Problem 1 Graph isomorphism

A famous problem in graph theory is identifying isomorphism between graphs. Put simply, two graphs are isomorphic if they have the same structure. More formally, we can permute the nodes (and preserve connecting edges) as to exactly transform one graph into the other.

For example, it is obvious that the left and middle graphs in Figure 5.18 are isomorphic. These two graphs are drawn a bit differently in the plane, but both have four nodes that are connected in the same manner. In contrast, the graph on the right is not isomorphic to these two: although it also has four nodes, these nodes are connected differently, and in particular the number of edges is different.

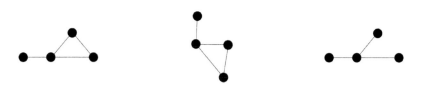

Figure 5.18. Only left and middle graphs are isomorphic

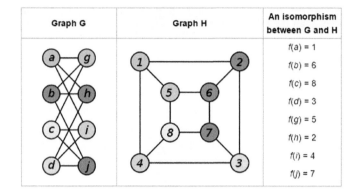

Figure 5.19. Isomorphic graphs. Source: Wikimedia Commons, created by Booyabazooka and used under Creative Commons Attribution-Share Alike 3.0 Unported license.

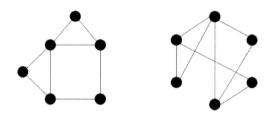

Figure 5.20. Are these graphs isomorphic?

The next example, in Figure 5.19, is a bit more involved. However, you should be able to convince yourself that these are indeed two graphs with an identical structure. We provide the mapping between the nodes of the two graphs (also color coded).

Interestingly, the notion of graph isomorphism is related to the notion (in chemistry) of chain isomerism (or skeletal isomerism).

Determining if two graphs are isomorphic is not a trivial task, but recent result by Laszlo Babai Babai (2015) indicates that the problem is not computationally very hard.

(a) Do you think the two graphs in Figure 5.20 are isomorphic? Try to find a sufficient explanation.

(b) Find an example for two graphs that are not isomorphic, though they have the same distribution of node degrees (that is, the same number of nodes with degree 0, 1, 2, 3, etc.). In particular, note that this requirement forces the two graphs to have the same number of nodes and edges.

(c) Conclude, from the previous section, whether "same degree distribution" is a necessary or a sufficient condition for two graphs to be isomorphic.

Problem 2 Hubs

In graphs, a **hub** is a node whose degree is exceptionally large compared to most other nodes. Hubs in biological networks may hint at an important role of the corresponding node, as it affects many other elements in the network directly. In directed graphs, we separate in-hubs from out-hubs, which are nodes of either a much higher in-degree or a much higher out-degree, respectively.

For simplicity, let us define a hub as a node with the *maximal* degree in the graph (for digraphs, maximal in-degree or out-degree). We note that the maximal degree need not be exceptionally larger than the other values).

Complete the following function, hub(G, inout). The first parameter is a directed graph, represented as a matrix of 0s and 1s (the graph is not weighted). The second parameter, inout, is either the string "in" or "out," and determines which type of hub to look for. The function will return a pair: the first value is the hub's degree, and the second value is the hub node index (recall that the index is an integer between 0 and len(G)-1).

```
1   def hub(G, inout):
2       ''' Find node with maximal in/out degree in G.
3           G is a non-weighted, digraph represented as a 0/1 matrix.
4           inout = "in" or "out"
5           Returns a pair (maximal degree, corresponding node) '''
6
7       max_deg = 0
8       hub_node = None
9       for i in range(len(G)):
10          if inout == "in":
11              ...
12              # variable deg will contain the in-degree of node i
13          elif inout == "out":
14              ...
15              # variable deg will contain the out-degree of node i
16          else: # inout must be "in" or "out"
17              return
18          if deg > max_deg:
19              max_deg = deg
20              hub_node = i
21      return max_deg, hub_node
```

Reference

Barabási, Albert-László, and Réka Albert. "Emergence of scaling in random networks." *Science* 286.5439 (1999): 509–512.

6   Shortest Paths and Breadth First Search

In this chapter, we introduce the notion of **graph algorithms**, which are basically algorithms working on graphs. There are many such algorithms, aimed at solving a wide range of problems. We will focus on one such problem – the **shortest paths problem**. This problem has several algorithms, under different constraints, that solve it. We will present the well-known **breadth first search** (BFS), algorithm that solves a simple version of that problem. This algorithm will be explained in detail and implemented in Python. We will conclude the chapter by mentioning additional common problems in graph theory.

6.1 The Shortest Paths Problem

A basic problem in graphs is the shortest paths problem. Suppose we have a graph (could be directed or undirected), and two nodes in it, v_1 and v_2. We wish to find the distance from v_1 to v_2, denoted by $dist(v_1, v_2)$, and in addition a path whose length is equal to the distance. In other words, we want to find a **shortest path** from v_1 to v_2.

Take for example the directed graph in Figure 6.1. The distance from node 0 to node 5 is 2. There are two distinct paths that give rise to this distance: (0,2,5) and (0,3,5) (recall that we denote a path as a sequence of the nodes along it). These are the shortest paths from node 0 to node 5. The shortest path from 0 to 4 is of length 3: (0,3,6,4).

This problem has many applications. Whenever you look for navigation instructions in your favorite navigation application, you basically solve an instance of this problem (albeit a more complicated version of it, that we will not discuss here). Other applications include routing an internet package to its destination using the fastest route, determining the minimal number of people that are in the path connecting two people in a social network, and, in a biological context, finding the fastest signaling path by which a signal is transmitted to its destination in a neural network.

We will deal with a simple variant of this problem: the graph will be non-weighted, which means that the length of a path is simply the number of edges in it. However, the algorithm that we will study works for both directed and undirected graphs.

6.2 Breadth First Search (BFS)

6.2.1 The Algorithm

The well-known BFS algorithm traverses a given graph, starting with an initial node. The algorithm "discovers" neighboring nodes in a wave-like manner: first, the immediate neighbors of the initial node are visited, and assigned distance 1 from the initial node. Then, the neighbors of each of these neighbors are visited. Among those, some have already been

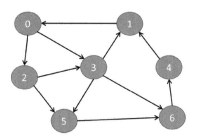

Figure 6.1. Shortest paths from node 0 to node 5 are 0→2→5 or 0→3→5, both of length 2. The distance from node 0 to node 4 is 3, and the corresponding shortest path is 0→3→6→4

discovered in the previous step. The others are assigned distance 2, and so on. At the end, every node that is reachable from the initial node is assigned the correct distance.

　　To maintain this wave-like behavior, the algorithm uses a list of the nodes currently waiting to be handled (that is, to have their neighbors explored). We call these nodes "active" nodes. At the beginning, only the initial node is active, and is the only node in that list. All other nodes are non-active, because they have not been reached yet. We refer to them as "unseen" nodes. When an unseen node is addressed for the first time, it enters the list of active pending nodes. We denote the oldest node in the list at each step of the algorithm by u.

　　Next, the neighbors of u are explored. If such a neighbor was not assigned a distance, it is assigned the distance assigned to u plus 1. The node u itself is removed from the list. Nodes removed from the list will never return to it. Thus, we call them "final."

　　Always taking the oldest element out of a set of elements is termed **FIFO**, which stands for "**First In First Out**." Just like real-life queues (at least those that are fair), the element with the longest waiting time (that is, the element that was inserted first among all current elements) is the next to be handled and removed from the queue. Indeed, computer scientists use the term 'queue' to describe such a FIFO-handled set of elements.

　　An example for the flow of the BFS algorithm is shown in Figure 6.2. The initial node is labeled s. For each step in the algorithm execution, the elements in the queue (denoted Q) are shown, including those that were previously removed from it. The queue is initialized with only s in it. In addition, the values of the distances computed along the algorithm execution are shown in a list denoted *dists*. At the beginning, all distances are equal to infinity, except for the initial node, s, whose distance (from itself) is obviously 0 (we can reach s from s through a path of length 0). Whenever unseen nodes are discovered, their distance from s is determined. Note that once a distance is stored in the list *dists*, it will never be changed. In other words, for any node that is reachable from s, its distance will be computed exactly once, and this is the final (and correct) value.

6.2.2 Implementation

We start with a description of the BFS algorithm in terms of **pseudo-code**. Such a description conveys the main details of the implementation but is still abstract enough to avoid some technical details. This is a recommended first step when implementing any algorithm.

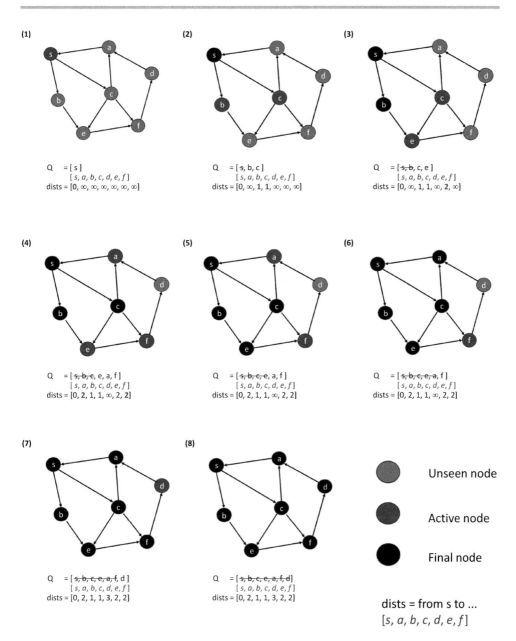

(1)

Q   = [s]
     [s, a, b, c, d, e, f]
dists = [0, ∞, ∞, ∞, ∞, ∞, ∞]

(2)

Q   = [s, b, c]
     [s, a, b, c, d, e, f]
dists = [0, ∞, 1, 1, ∞, ∞, ∞]

(3)

Q   = [s, b, c, e]
     [s, a, b, c, d, e, f]
dists = [0, ∞, 1, 1, ∞, 2, ∞]

(4)

Q   = [s, b, c, e, a, f]
     [s, a, b, c, d, e, f]
dists = [0, 2, 1, 1, ∞, 2, 2]

(5)

Q   = [s, b, c, e, a, f]
     [s, a, b, c, d, e, f]
dists = [0, 2, 1, 1, ∞, 2, 2]

(6)

Q   = [s, b, c, e, a, f]
     [s, a, b, c, d, e, f]
dists = [0, 2, 1, 1, ∞, 2, 2]

(7)

Q   = [s, b, c, e, a, f, d]
     [s, a, b, c, d, e, f]
dists = [0, 2, 1, 1, 3, 2, 2]

(8)

Q   = [s, b, c, e, a, f, d]
     [s, a, b, c, d, e, f]
dists = [0, 2, 1, 1, 3, 2, 2]

○ Unseen node

● Active node

● Final node

dists = from s to ...
[s, a, b, c, d, e, f]

Figure 6.2. An example for the flow of the BFS algorithm

BFS (*G*, *source*):

Initialization:

1. *dists* = 0 for *source*, and infinity for all other nodes in *G*
2. *Q* = [*source*], the only node we "discovered" so far

Main Loop:

3. while there are nodes in Q to handle:
   3.1 $u \leftarrow$ the head of Q (oldest node)
   3.2 remove u from Q (we are done with it)
   3.3 for every neighbor v of u in G:
        3.3.1 if v is seen for the first time (i.e., its distance is infinity):
               3.3.1.1 update $dist[v]$ to be $dist[u] +1$
               3.3.1.2 insert v into Q

Termination (Q is empty):

4. return *dists*

    In Python, infinity can be represented by the expression `float('Inf')`. This expression "behaves" like infinity; that is, it is larger than any number:

```
>>> print(float('Inf'))
Inf
>>> inf = float('Inf')
>>> inf > 100**100
True
>>> inf < 100*100
False
```

Very convenient, isn't it?
    We are finally ready to see the implementation of BFS in Python:

```python
1 def bfs(G, source):
2     inf = float('Inf')
3     n = len(G) # number of rows in matrix = number of nodes
4
5     # Initialization
6     dists = [inf]*n
7     dists[source] = 0
8
9     Q = [source] # Queue of "active" nodes
10
11    # Main loop of the algorithm
12    while len(Q) != 0:        # while there are still nodes to handle
13        u = Q.pop(0)          # get and remove head of Q
14        for v in range(n):  # go over all neighbors v of u
15            if G[u][v] == 1 and dists[v] == inf: # first time in v
16                dists[v] = dists[u] + 1          # update v's dist
17                Q.append(v)                       # v enters Q
18
19    # Termination, Q is empty
20    return dists
```

As a specific example, the input to our algorithm will be the directed graph in Figure 6.3, in its matrix representation.

```
G = [ [0, 0, 1, 1, 0, 0, 0],
      [1, 0, 0, 0, 0, 0, 0],
      [0, 0, 0, 1, 0, 1, 0],
      [0, 1, 0, 0, 0, 1, 1],
      [0, 1, 0, 0, 0, 0, 0],
      [0, 0, 0, 0, 0, 0, 1],
      [0, 0, 0, 0, 1, 0, 0], ]
```

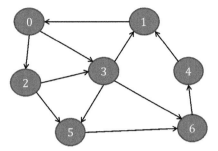

Figure 6.3. An example graph for the illustration of BFS

Here are some execution examples. Let us first run BFS from node 0:

```
>>> G = ...    <= the matrix from Figure 6.3
>>> s = 0
>>> dists = bfs(G, 0)
>>> print(dists)
[0, 2, 1, 1, 3, 2, 2]
```

Code explained

In line 1, we initialize `inf` to be used as infinity (a node where no path from the source was established yet). In line 2, `len(G)` is used. This is the number of elements in `G`. Since each element in `G` is a list, `len(G)` is the number of such lists, or the number of rows in the matrix representing the graph. The lines after that initialize the variable `dists` and the queue `Q`, as explained earlier.

In the main loop, as long as `Q` is not empty, we pop the oldest element, in line 13 (the parameter for the pop method of class `list` denotes the index of the element to remove). We then go over all the nodes `v` in the graph, and if there is an edge from `u` to `v` (`G[u][v] ==1`) and `v` is discovered for the first time (`dists[v] == inf`), we update its distance from the origin to be `u`'s distance plus 1, and insert `v` into the end of the queue.

Exercise 1

Follow the execution of BFS from the source node 3. What is the returned value of the function?

Exercise 2

If a node is not reachable from the source, what will its distance be at the end of the algorithm's execution?

> ⊘   Exercise 3
>
> Another variation of the shortest paths problem is that we may require the path to go through a particular node. Can you suggest a simple solution to this version, without changing the inner details of the BFS algorithm?

6.2.3 Reconstructing the Shortest Paths

So far, we have only computed the lengths of the shortest paths, or distances, from some source node to all other nodes in the graphs. Our last step in the implementation of BFS will be to generate the shortest paths themselves. For a given run of BFS from a given source node s, and for any target node t that is reachable from the source, we will construct a shortest path from s to t *backwards*.

The node that is located just before t on the shortest path from s must be the node that "discovered" t in the BFS execution (namely the node that set t's distance), so there is no shorter path to t. To that end, we will add to our implementation an additional list, called prevs. It will store, for every node in the graph, the node which "discovered" it for the first time. We will then be able to repeatedly go backwards from t to the previous node on the path, until we reach s.

The lines we added to our implementation are highlighted:

```
1   def bfs(G, source):
2       inf = float('Inf')
3       n = len(G) # number of rows in matrix = number of nodes
4
5       # Initialization
6       dists = [inf]*n
7       dists[source] = 0
8       prev = [-1]*n   # initial values of -1 denote node not discovered yet
9
10      Q = [source]  # Queue of "active" nodes
11
12      # Main loop of the algorithm
13      while len(Q) != 0:  # while there are still nodes to handle
14          u = Q.pop(0) # get and remove head of Q
15          for v in range(n):  # go over all neighbors v of u
16              if G[u][v] ==1 and dists[v] == inf:  # v seen first time
17                  dists[v] = dists[u] + 1         # update v's dist
18                  prevs[v] = u                    # u is v's previous
19                  Q.append(v)                     # v enters Q
20
21      # Termination, Q is empty
22      return dists, prevs
```

```
>>> G = ... <= the matrix from Figure 6.3
>>> s = 0
>>> dists, prevs = bfs(G, 0)
>>> print(dists)
[0, 2, 1, 1, 3, 2, 2]
>>> print(prevs)
[-1, 3, 0, 0, 6, 2, 3]
```

Note that we initialized the list `prevs` by −1, and for the source node this value remained unchanged.

Exercise 4

If a node is not reachable from the source s, what will its corresponding value in the list `prevs` be at the end of the algorithm's execution?

The function `shortest_path` receives the graph `G`, a `source` node and a `target` node. It returns the path between the source and the target, which is guaranteed to be the shortest one:

```
1   def shortest_path(G, source, target):
2       ''' Returns the shortest path in G from source to target.
3           Uses BFS's output.
4           G is non-weighted, directed graph as a binary matrix.
5           The output is a list of the nodes in the path '''
6       dists, prevs = bfs(G, source) # dists will not be used
7       if prevs[target] == -1:   # target not reachable from source
8           return []
9       path = [target]
10      v = target
11      while v != source:
12          v = prevs[v]      # previous node in the path
13          path = [v] + path   # append v at left end of path
14      return path
```

Code explained

The function runs `bfs` first to get the list `prevs`. This list will be used to extract the path from source to target, backwards (from last to first).

First, if `prevs[target]` remained −1 and not changed along the algorithm's execution, then `target` is not reachable from `source`. An empty path is returned in this case, represented by an empty list. Now, we initialize `path` to contain the last node in the path, which is the `target` node. At each iteration of the loop, we check if the current node `v` in the path (built backwards) is already the source node. If it isn't yet so, we move a step "backwards" to the source node, by replacing v with its previous node (`prevs[v]`), and add that previous node to `path` from the left. Once we have reached the source node this way, we have finished constructing the path.

Exercise 5

Run the function `shortest_path` to find the shortest path from node 0 (source) to node 4 (target) in Figure 6.3.

6.3 Related Biologically Relevant Metrics

In a paper published at the *Journal of Biomedicine and Biotechnology*, titled "High-betweenness proteins in the yeast protein" [see reference list], the authors use the notion of **betweenness** of a node. This is defined as the number of shortest paths that go through each node. They studied a yeast protein interaction (PPI) network of 4721 proteins (nodes) with 15,210 interactions (edges). Intuitively, a protein located on the shortest path between two other proteins has the most influence over the "information transfer" between them. Therefore, a high betweenness of a node may indicate its centrality in the global network behavior.

The paper suggests that high-betweenness low-degree proteins in the yeast PPI network are the most important to the network function. Figure 6.4 shows an example for a high-betweenness low-degree node (the red node). Such a node is likely to be a "bridge" between two or more clusters in the network.

Another simple measure is the **characteristic path length**: the average length of all shortest paths (between any two nodes). Intuitively, this measure is correlated with a graph's "spread-out". It can easily be computed by averaging the lengths of the shortest paths between each pair of nodes in the graph (note that the graph must be connected, otherwise we have infinite distances). It is reasonable to expect very different biological functions to differ in this measure. Another important metric is the maximal length over all shortest paths, termed the graph's **diameter**. In other words, the diameter is the distance between the farthest nodes. Both metrics are examples of global network properties.

Figure 6.4. The red node in this graph has a high-betweenness and low-degree values

---

> ### Exercise 6
>
> Write two functions `avg_dist(G)` and `diameter(G)`, which receive an unweighted graph G (directed or undirected), represented as a binary matrix, and return the average and maximal (corresponding) length of the shortest paths in G.

Reflection

In this chapter, we explored the BFS algorithm, which is used to solve a simple version of the shortest paths problem, for unweighted graphs. Shortest path graph algorithms are important in web applications that recommend efficient paths in traffic. Of course, such algorithms do not use a static graph, but are dynamic – the underlying graphs change over time.

There are several variants for the shortest paths problem, in terms of the types of graphs they handle, or additional constraints they include. Perhaps the most common variant is that of a weighted graph. The length of a path in a weighted graph is defined as the sum of weights along it. The BFS algorithm does not take edges' weights into account. There are several algorithms that handle this variant. A notable one is **Dijkstra's shortest paths algorithm**. It is named after Edsger W. Dijkstra, an influential, Dutch computer scientist (1930–2002). Dijkstra's algorithm handles weighted graphs, yet it is restricted to the case where all the edges have non-negative weights. It works similarly to the BFS algorithm, but nodes are extracted from the queue in a different manner, not in a simple FIFO manner.

Dijkstra's algorithm does not guarantee to work properly when the graph contains negative weights. However, there are other algorithms that handle such a case. A special case is if the graph contains a negative cycle (a cycle for which the sum of weights along it is negative). In such a case, the shortest paths problem is not well defined as it is possible to travel this cycle endlessly and reduce the weight of the path between any two nodes on the cycle infinitely. We remark that there are efficient algorithms that check if a graph contains such negative cycles.

BFS can be thought of as a way to traverse the nodes of a graph in a particular order. **Depth first search** (DFS) is another way to do so. DFS starts with a given node in the graph and explores paths as far as possible before backtracking to another branch. For example, in the graph in Figure 6.3, assuming smaller node indices are selected first, DFS from node 0 will explore the branch of nodes 2, 3, and 1, then backtrack to 3, and proceed to nodes 5, 6, and 4. The time complexity of the BFS and DFS algorithm has the same order of magnitude. The choice between them is usually based on properties of the traversal order, which yield different applications for each one.

There are numerous other algorithms on graphs, which solve various additional problems, some of biological relevance. For example, we may want to find the largest clique in a graph, the cluster with highest densities, the cycles of various sizes, the smallest subset of the edges whose removal disconnects the graph, or to compute the maximal flow that can be pushed through the graph from a given source to a given sink, etc.

Further Reading

Joy MP, Brock A, Ingber DE, Huang S. "High-betweenness proteins in the yeast protein interaction network." *J Biomed Biotechnol*. 2005.2 (2005): 96–103. doi:10.1155/JBB.2005.96.

Ma'ayan A, Jenkins SL, Neves S, et al. "Formation of regulatory patterns during signal propagation in a Mammalian cellular network." *Science*. 309.5737 (2005): 1078–1083. doi:10.1126/science.1108876.

Challenge Yourself

Problem 1 Checking connectivity via BFS (in undirected graphs)

1. How can the BFS algorithm be used to check if a given unweighted graph is connected or not?
2. Write a Python function `connected(G)`, that checks if `G` is connected.

Problem 2 The mammalian cellular network case study

In a paper published in the *Science* magazine, titled "Formation of regulatory patterns during signal propagation in a mammalian cellular network", a network of 545 components (nodes) and 1259 interactions (edges) was studied. This network is a simplified representation of the signaling pathways and cellular machines in the hippocampal CA1 neuron. The network includes extracellular ligands, linked to membrane proteins and receptors, from which the signals flow downstream until they reach various machinery components, such as translation, secretion, and motility. The study analyzed the location of various recurring patterns, called motifs, on the shortest paths from ligands to machinery components.

  The network used in the paper can be downloaded from the paper's supplementary material, and appears on the website of this book. You will also find there a function called `CA1file2mat`, which transforms the data file into a binary matrix of the form we used in this chapter to specify graphs.

1. Run this function on the data file, to generate the matrix representing the CA1 network:

```
>>> G,nodes = CA1file2mat("./CA1.txt")
```

    Now the variable `nodes` contains the list of nodes' names, and `G` contains the matrix. Do not try to display it on your computer screen, as most likely your editor will get stuck – this is a matrix of size 545-by-545.
2. Find the shortest path from the node "GLUTAMATE," representing an extracellular ligand activating a signal, and the node "CREB" representing the transcription factor cyclic-AMP response element binding protein.
3. The first edge in that path is from "GLUTAMATE" to "NMDAR."Disconnect these nodes by removing this edge from `G` and repeat the previous question. Did the network find a way to bypass this removed connection?
4. Can you compute the diameter of this network?

7 Simulation of Regulatory Networks

Many biological systems can be explored by computer **simulation**: an imitation of the behavior of the system over time, done by running a computer program. Computer simulations are also termed *in-silico* experiments, paraphrasing the terms *in-vitro* and *in-vivo*. Computer simulation in biology can be used to replace or complement some tedious and costly lab experiments. It enables conducting numerous "experiments" under various conditions, at a scale that is infeasible experimentally.

Running a computer simulation requires constructing a **mathematical model** of the biological system – some formal representation of the system using mathematical equations and algorithms to process the data. In other words, a mathematical model is a formal abstraction of reality. A valuable model should capture the relevant aspects, or essence of the system, with the appropriate level of detail.

Another important classification of models divides them into **continuous** and **discrete** models. In continuous models, either the time or the states of the system components (expression levels, half-time constants, etc.), or both, are described by real numbers. Such networks may provide detailed descriptions of the biological systems under study. In particular, continuous models often consist of some sort of differential equations. On the negative side, they often require a very detailed knowledge of quantitative biological data, that is, parameters such as concentrations and kinetic constants, whereas such knowledge may be unavailable.

Discrete models limit the time steps or the states of components, or both, to the integer $(0,1,2,\ldots)$. Despite the lower granularity in the system's description, discrete models provide a satisfactory alternative in many cases: they tend to be more intuitive, often do not require full knowledge of the system as a starting point, and may be more computationally efficient.

Boolean models are a special case of discrete models. In Boolean models, variables are limited to only two values such as 1 and 0. If, for example, these values represent expression levels, we relate to entities in the system as merely "active" or "non-active." Although Boolean models provide the lowest possible resolution, they have proved valuable in various research studies, mainly for qualitative exploration.

In this chapter, we present a discrete computational model for the simulation of biological regulatory networks. The model is discrete both in terms of the states of entities (nodes) and in terms of time progress ("clock ticks"). We start by describing the model non-formally using a toy example. We then give a precise definition of the model, followed by its implementation in Python to allow its application to biological data. Finally, we describe several examples for studies conducted with it. As these examples show, this model is highly valuable for the analysis of systems that are devoid of consistent quantitative data, as is often the case in biological networks.

7.1 A Non-Formal Description of the Model

7.1.1 The General Setting

The model we are going to describe here is a **state graph**. This means that a biological network or pathway is represented by a graph (nodes connected by edges), and that this graph has a state, as will be explained below. Nodes in the graph represent either physical entities, such as mRNA, proteins, and nutrients, or events, such as cell division and temperature change. Edges represent regulation interactions, and have weights: the weight of an edge is determined by the strength of the interaction it represents. Positive weights stand for **activation** and negative weights for repression or **inhibition**.

Each node in the graph has a **state**, taken from a fixed, discrete range of integers, such as 0,1,2,...,9 (the minimal state is always 0, while the maximal state can be defined to be different than 9). This range provides a resolution of 10 different levels. For physical entities, the state of a node represents its relative expression level, concentration, or activity: state 0 stands for lack of activity, or absence of the entity represented by the node, while state 9 reflects full activity or maximal possible expression level. For abstract entities, the meaning may vary according to the context, e.g., progress of a cellular event and level of temperature.

Biological data regarding the network's initial conditions are used to set an **initial state** for each node in the network. These states are used as the starting point for the simulation, and may change along the course of it as a result of activation and inhibition effects by neighboring nodes in a manner to be explained soon. The simulation consists of discrete time steps (step 0, step 1, step 2, and so on). Once all the nodes' states do not change anymore, or are "trapped" in a cycle (examples later), the simulation ends.

This is basically the whole story, the rest is just details ... but for now let us turn to a simple toy example, and use it to explain how the simulation is conducted.

7.1.2 Toy Network

Our initial example is a simple network of three nodes: A, B, and C (see Figure 7.1). Node A activates node B, node B activates node C, and node C inhibits node A. The weight of the two activation edges is set to +1, while the inhibition edge is set to −1. In Figure 7.1, the edges' weights are not shown, but activation edges are green arrows and inhibition edges are red lines with a bar. The initial state of nodes A is 9 and that of B and C is 0 (initial states are

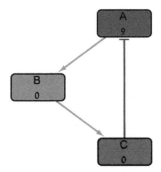

Figure 7.1. Toy network containing three nodes and three edges. Green edges represent activation, red edges – inhibition.

written below a node's name in Figure 7.1). We denote the network state as a sequence of all nodes' states $(A,B,C) = (9,0,0)$.

The **effect** of some node i on some node j is defined as the product of the state of node i and the weight of the edge from i to j. Note that the larger the state of node i, the larger its effect on node j, and this effect is also proportional to the edge's weight. For example, the effect of node A on node B is $9 \cdot (+1) = +9$, while the effect of node B on node C is $0 \cdot (+1) = 0$ (that is, it has no effect), as is the effect of node C on node A $(0 \cdot (-1) = 0)$.

7.1.3 Simulation

Let us now explain the process of simulation of this network in detail. For each node, we sum up the effects of its incoming edges. Note that in this simple network, each node has only a single incoming edge, so the summation of effect on each node consists of a single value. Later on, we will see more complex examples in which nodes will be affected by multiple neighbors. The sum of effects on each node, or the "net" effect, will determine how its state will be updated in the next simulation step. If this sum is positive, meaning that in total the activation overpowers the inhibition, the state will increase by 1; if it is negative, meaning that inhibition effects exceed activation, the state will decrease by 1; and if the sum is 0, we have a "tie," and the state will remain unchanged. This computation is done on every node in the network at each time step.

Node A, for example, begins at state 9. The only edge affecting this node is an inhibitory edge with weight -1 from node C. However, in time step 0, node C is at state 0, and its effect on node A is 0. Therefore, the state of node A remains unchanged. At the same time (simulation step 0), node B is activated by a "net" effect of $+9$ (due to node A). So in step 1, the state of node B will increase to 1. Node C will remain in state 0 (make sure you understand why). So in simulation step 1, the network state will be $(A,B,C) = (9,1,0)$.

Exercise 1

Explain why the network state in step 2 will be $(A,B,C) = (9,2,1)$.

In step 2, the state of node C is not 0 anymore. Therefore, it starts to inhibit node A. In fact, node B continues to increase step by step (up to state 9 in simulation step 9), node C follows it (up to state 9 in step 10), while node A, which started to decrease in step 2, reaches state 0 soon after (at step 11). The states of the three nodes along the simulation are summarized in Table 7.1, and appear graphically in Figure 7.2(a).

What happens from simulation step 11 onwards? In step 12, node states remain unchanged, so the network state is identical to that in step 11, $(A,B,C) = (0,9,9)$. Once the network state repeats itself in two consecutive simulation steps, it is "trapped" in a **steady state**, or **fixed point**. From this step on, the states of nodes will not change anymore. This is one way by which a simulation can terminate (the other – a simulation cycle – will be shown later).

What would happen if we changed the initial state of node C to, say, 5? Figure 7.2(b) shows the results. Note how node A immediately starts decreasing, reaching state 0 as early as step 9.

Table 7.1. *A simulation run*

	Step 0	Step 1	Step 2	Step 3	Step 4	Step 5	Step 6	Step 7	Step 8	Step 9	Step 10	Step 11	Step 12
A	9	9	9	8	7	6	5	4	3	2	1	0	0
B	0	1	2	3	4	5	6	7	8	9	9	9	9
C	0	0	1	2	3	4	5	6	7	8	9	9	9

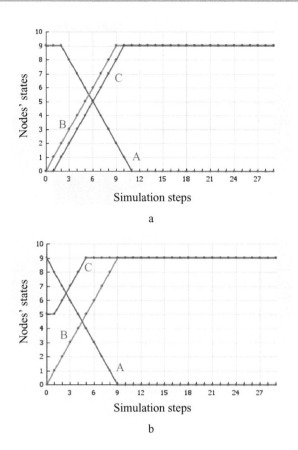

Figure 7.2. (a) Simulation results of the network from Table 7.1, when node C starts from state 0 (b) Simulation results when node C starts from state 5

Note that the order by which we update states within each step does not matter. We compute all the new states of all nodes based on the current state of the network, before making the update and moving on to the next step. We call such a model **synchronous**. A variation of this model could be **asynchronous**, meaning that while each node is updated once per simulation step, a change in a node's state takes place immediately, and may affect the update of other nodes. The order of the update of other nodes will then matter.

7.1.4 Negative Loops

A negative loop is an edge from a node to itself (a loop) with a negative weight. Such loops are often used in regulation network models to represent the natural degradation of a substance: even without an inhibitory effect, the substance (e.g., protein) expression level will decrease over time due to its own natural degradation. It is not necessarily the case that the node inhibits its own activity, but this is merely a way to model the fact that a node's state will decrease independently of other effects.

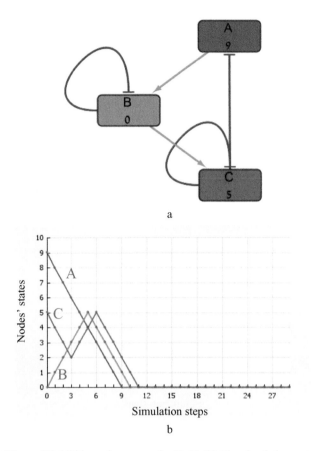

Figure 7.3. (a) Adding self-inhibition edges on nodes B, C. (b) The simulation output.

As an example, let us add to our toy network self-inhibitory edges on nodes B and C with weight −1. Figure 7.3 shows the simulation results of this modified network.

Exercise 2

In Figure 7.3, the state of node C decreased until step 3, but then started increasing until step 6. Explain why this happened. Why did it start decreasing again?

7.1.5 Simulation Cycles

We mentioned before that a simulation ends when either the system has reached a steady state or when it has reached a **cycle**. We saw examples for the first case, let us now see an example for the other one. (In fact, one can relate to a steady state as a simulation cycle of length 0. We use the term **attractor** to denote the steady network state, or the cycle of network states, which the simulation converges to.)

> ### Exercise 3
>
> As an example for a simulation cycle, consider the simple network in Figure 7.4. The initial network state is $(X,Y) = (1,0)$. What will the network state be at step 1? Step 2? Step 3? Assume edges' weights are +1 (green) or −1 (red).
>
>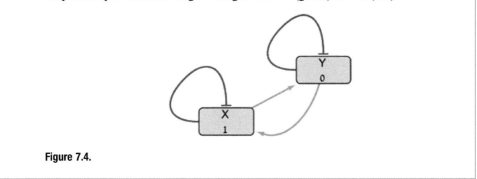
>
> **Figure 7.4.**

Back to our toy example, we will now add to it an additional node, denoted "signal" that represents a biological or chemical entity that constantly activates the system (see Figure 7.5(a)). The initial state of this node will be set to 1, and it will activate node A, with an edge of weight +1. Note that since the signal node is not regulated by any other node – that is, it has no incoming edges – its state will remain unchanged. The simulation of this modified network is shown in Figure 7.5(b) – a cyclical expression pattern in which A, B, and C go up and down sequentially. We can terminate the simulation when we identify such a cyclical pattern, because the system will continue to behave like this forever.

7.2 Model Definition

This section presents a more formal description of our model. One advantage of writing down formal mathematical descriptions of models is that later, when you turn to implement them in code, things become much easier. The mathematical definitions often naturally translate into code very smoothly. Another advantage is that formal definitions enforce you to think of all the fine details, special cases, etc.

Our model is a directed graph, and we assume nodes are indexed by $0,1,\ldots,n-1$ (the number of nodes is denoted n). Each node has a state taken from the range $\{0,1,\ldots,U_i\}$, where U_i is the maximal state for node i ($U_i > 0$). For example, if for some node $U_i = 9$, then its states are limited to the range $0,1,2,\ldots,9$. Although U_i may vary from node to node, it may sometimes be convenient to define a common upper bound for all nodes, U (for example, $U = 9$, that is, all nodes' states range between 0 and 9). Time in the model consists of discrete steps, or "clock ticks," $t = 0,1,2,\ldots$. The state of node i at time step t is denoted $s_i(t)$. Each node is set to have an initial state $s_i(0)$. The state of the whole network at time step t is defined as the sequence of the states of the nodes, $(s_0(t), s_1(t), \ldots, s_{n-1}(t))$. Therefore, the initial network state is $(s_0(0), s_1(0), \ldots, s_{n-1}(0))$.

How can we express the "net" effect on some node i at time step t in a formal manner? Recall that this effect is the sum of individual effects of incoming edges. The effect of node j on node i is the product between the state of node j in the current time step, $s_j(t)$, and the weight of

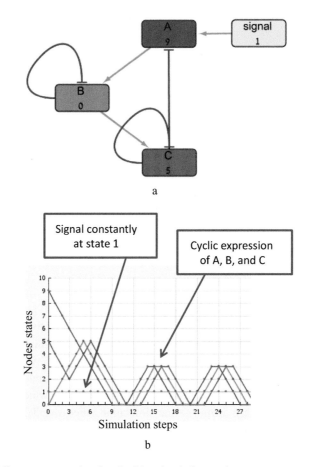

a

b

Figure 7.5. (a) Adding a constant signal node (b) a simulation cycle

the edge (j, i). Let us denote the latter $w(j, i)$, and so the product is $s_j(t) \cdot w(j, i)$. This product needs to be summed over all nodes j in the network for which there is an edge (j, i).

To make things simpler, we can relate to missing edges as edges with weight 0. This way, if there is no edge (j, i), then $s_j(t) \cdot w(j, i) = 0$, and thus will have no effect on the overall sum anyway. This allows us to simply compute the sum of $s_j(t) \cdot w(j, i)$ over all nodes j in the network (without separation between existing and missing edges). So the total effect on node i at any time step t is expressed by the following sum:

$$s_0(t) \cdot w(0, i) + s_1(t) \cdot w(1, i) + s_2(t) \cdot w(2, i) + \cdots + s_{n-1}(t) \cdot w(n-1, i)$$

Instead of writing this long series, we can use the summation notation $\sum$:

$$\sum_{j=0}^{n-1} s_j(t) \cdot w(j, i)$$

Also, instead of writing this complex expression every time we use it, we can denote it as, say, $\sigma_i(t)$ (σ is the Greek letter sigma). This is merely a matter of convenience, but it is something mathematicians do a lot in order to avoid very complex expressions. So the total effect on node i at time step t is:

$$\sigma_i(t) = \sum_{j=0}^{n-1} s_j(t) \cdot w(j, i)$$

Exercise 4

Suppose the nodes in Figure 7.5(a) are indexed as follows: $i = 0, 1, 2, 3$ for nodes A, B, and C, signal. Compute $\sigma_i(0)$, for every i.

A simulation starts with the initial network state $(s_0(0), s_1(0), \ldots, s_{n-1}(0))$, and consists of the repeated application of a transition function in a synchronous manner (all nodes are affected "at once"). The new state of node i in the next time step, $s_i(t + 1)$, will remain unchanged, if $\sigma_i(t) = 0$; it will increase by 1, if $\sigma_i(t)$ is positive; and it will decrease by 1 if $\sigma_i(t)$ is negative. Formally, when $\sigma_i(t) = 0$, then $s_i(t + 1) = s_i(t)$. When $\sigma_i(t) > 0$, then $s_i(t + 1) = s_i(t) + 1$ if, and when $\sigma_i(t) < 0$, we get $s_i(t + 1) = s_i(t) - 1$. We can put all this together as the following:

$$s_i(t + 1) = \begin{cases} s_i(t) + 1 & \text{if } \sigma_i(t) > 0 \\ s_i(t) - 1 & \text{if } \sigma_i(t) < 0 \\ s_i(t) & \text{else} \end{cases}$$

Exercise 5

Suppose the nodes in Figure 7.5(a) are indexed as follows: $i = 0, 1, 2, 3$ for nodes A, B, C, signal. Compute $s_i(1)$, for every i.

There is one last thing to take care of: overflow (states exceeding U_i) and underflow (states decreasing below 0). If $\sigma_i(t)$ is positive, the state of node i in the next time step $t + 1$ will increase by 1, but should not exceed the maximal possible state for that node U_i. So if $s_i(t) + 1$ exceeds U_i, the new state will be U_i. In other words, the new state will be the minimum between $s_i(t) + 1$ and U_i. Similarly, when $\sigma_i(t)$ is negative, the new state will be the maximum between $s_i(t) - 1$ and 0. We can use the mathematical functions *min* and *max* in our formula to express this in the following way:

$$s_i(t + 1) = \begin{cases} min(U_i, s_i(t) + 1) & \text{if } \sigma_i(t) > 0 \\ max(0, s_i(t) - 1) & \text{if } \sigma_i(t) < 0 \\ s_i(t) & \text{else} \end{cases}$$

Readers who feel really comfortable with mathematical expression may enjoy the following, even more compact version of the above formula:

$$s_i(t + 1) = max\,(0,\, min\,(U_i, s_i(t) + sign(\sigma_i(t))))$$

where $sign(\sigma_i(t))$ equals $+1$ when $\sigma_i(t)$ is positive, -1 when it is negative, and 0 when it is 0. The sign function is a common notation in mathematics.

Recall that a simulation will end when it converges to an attractor: either a steady state (two consecutive identical network states), or a loop of network states.

7.3 Implementation of the Model in Python

In the previous section, we formulated the model in concise, mathematical terms. This will make it easier to "translate" the model into a computer program, which we will do in this section.

7.3.1 Representation of the Network and Simulation

A network can be represented by a matrix with edges' weights. Entry (i,j) will hold the weight of the edge from node i to node j. For example, our toy network from Figure 7.3 will be represented as follows (the rows, corresponding to nodes, are ordered arbitrarily):

```
1 G = [    [0, 1, 0],      # node A
2          [0,-1, 1],      # node B
3          [-1, 0,-1]  ]   # node C
```

The state of the network at a specific time step will be represented as a list of nodes' states. For example, [9,0,5] is the initial network state of the simulation in Figure 7.3. A simulation is represented by a sequence of lists, one for each step.

7.3.2 The Model's "Engine"

The next three functions implement the model and its simulation. For simplicity, we will assume all states of the nodes have a common upper bound, stored in a variable U. This variable is defined as a global variable, that is, outside the functions, so all the functions recognize it without it being transfered as a parameter. This is merely a choice of convenience.

```
1 U = 9 # common state upper bound
```

The function `update_node` computes the next state of a given node. It receives the network as a matrix G, a list of current nodes' states, and the index of the node we want to update.

```
1 def update_node(G, U, curr_states, i):
2     ''' G - a network (weighted matrix, G[v][u] is weight from v to u)
3         U - maximal state
4         curr_states - a list nodes' current states
5         i - index of node to update (between 0 and len(G)-1)
6         apply transition function to node i and return its new state
7     '''
8     n = len(curr_states)
```

```
9      s = 0
10
11     for j in range(n):
12         s += G[j][i] * curr_states[j]
13     if s > 0:
14         new = min(U, curr_states[i] +1)
15     elif s < 0:
16         new = max(0, curr_states[i] -1)
17     else:
18         new = curr_states[i]
19
20     return new
```

Code explained

The "for" loop in lines 10 and 11 iterates over all node indices j, and computes the "net" effect on node i by summing up all effects caused by each node j. The rest of the function separates between a positive, negative, and zero net effect, and returns the appropriate new state of node i.

```
>>> update_node(G, 9, [9,0,5], 0)  <= update node A
8
>>> update_node(G, 9, [9,0,5], 1)  <= update node B
1
>>> update_node(G, 9, [9,0,5], 2)  <= update node C
4
```

Next, the function step iterates over all the network nodes and calls update_node:

```
1  def step(G, U, curr_states):
2      ''' G - a network (weighted matrix, G[v][u] is weight from v to u)
3          U - maximal state
4          curr_states - a list nodes' current states
5          Return the list of states in the next time step.
6          Apply transition function to all nodes synchronously
7      '''
8      n = len(curr_states)
9      new_states = [None]*n
10
11     for i in range(n):
12         new_states[i] = update_node(G, U, curr_states, i)
13
14     return new_states
```

Code explained

All the new states of nodes will be stored in the list `new_states`. It is initialized as a list of size n, the number of nodes in the graph. The value `None` is stored temporarily in each index. The loop calls the function `update_node` for each node `i` in the graph, and stores the next state of node `i` at that index.

```
>>> step(G, 9, [9,0,5])
[8, 1, 4]
>>> step(G, 9, [8,1,4])
[7, 2, 3]
```

Finally, we have the function `run`, which calls the function `update` repeatedly. We show here a non-final version of this function, which assumes that the simulation converges to a steady state (and not to a cycle. An exercise at the end of this chapter will guide you in adding the detection and printout of simulation cycles). The function `run` will return the trajectory of the simulation, that is, a list of network states, one for each simulation step, starting from the initial one.

```
1   def run(G, U, init_states):
2       ''' G - a network (weighted matrix, G[v][u] is weight from v to u)
3           U - maximal state
4           init_states - list of initial states
5           Run simulation until steady state (TODO: cycles)
6           Return simulation trajectory
7       '''
8       trajectory = [init_states]
9       curr_states = init_states
10      next_states = step(G, U, curr_states)
11
12      while next_states != curr_states: # not a steady state
13          trajectory += [next_states]
14          curr_states = next_states
15          next_states = step(G, U, next_states)
16
17      trajectory += [next_states]
18      return trajectory
```

Code explained

The variable `trajectory` will store the sequence of network states along the convergence to a steady state when this function ends (see running example below). So the number of elements in this variable is the same as the length of the simulation (until convergence).

At the beginning, `trajectory` contains only the initial network state. The variables `curr_states` and `next_states` in lines 9 and 10 store the current and next network states (using the step function we saw before). Now, as long as these two are not identical (when they are, a steady state has been reached), we move forward in the simulation one step at a time (lines 14 and 15), not forgetting to add the new network state to the `trajectory` (line 13).

When we reach a steady state, the loop ends and we are ready to return the `trajectory`. Just before that, we add the last network state to the trajectory in line 17. This means that the last two elements in the trajectory are identical lists of the states of the nodes. This will turn out to be convenient later, when we need to identify whether a specific trajectory represents a simulation that ended in a steady state or a loop.

```
>>> run(G, 9, [9,0,5])
[[9, 0, 5], [8, 1, 4], [7, 2, 3], [6, 3, 2], [5, 4, 3], [4, 5, 4], [3,
4, 5], [2, 3, 4], [1, 2, 3], [0, 1, 2], [0, 0, 1], [0, 0, 0], [0, 0, 0]]
```

This is not a very clear output, is it? How about a nice function, `pretty_print`, which takes this output, a list of inner lists representing network states, and prints them more clearly?

```
1 def pretty_print(trajectory):
2     ''' Get simulation trajectory (list of lists, one per step)
3         and print it nicely
4     '''
5     print("step:  network state:")
6     for i in range(len(trajectory)):
7         print(i, ":", trajectory[i])
```

```
>>> pretty_print(run(G, 9, [9,0,5]))
step:  network state:
0 : [9, 0, 5]
1 : [8, 1, 4]
2 : [7, 2, 3]
3 : [6, 3, 2]
4 : [5, 4, 3]
5 : [4, 5, 4]
6 : [3, 4, 5]
7 : [2, 3, 4]
8 : [1, 2, 3]
9 : [0, 1, 2]
10 : [0, 0, 1]
11 : [0, 0, 0]
12 : [0, 0, 0]
```

7.3.3 Gathering Attractors Statistics

Suppose we wanted to examine which steady states are reached by a given network, when all the combinations of initial states are considered. For example, if we have a network with two nodes, and each can assume the initial state between 0 and 9, then we have $10^2 = 100$ possible combinations of initial states: [0,0], [0,1], [0,2], ..., [0,8], [0,9], [1,0], [1,1], [1,2], ..., [1,8], [1,9], [2,0], [2,1],...

We can run a simulation for each such initial state, and see which steady state is reached (assuming we do not enter a cycle). Then, we may want to see the "big picture", that is, how many network initial states converged to each of the steady states. It is possible for example that all 100 network initial states converged to the same steady state, or that 90 of them converged to one steady state, six additional ones to another, and the four remaining to a third steady state. In this example, we have one "main attractor", and two additional, smaller ones.

Gathering such statistics enables a higher-level study of a network, and may reveal global properties. For example, if we get one main attractor as in the example above, we may conclude that the network is relatively resistant to perturbations in initial states of various nodes. This can be interpreted as robustness of the system being modeled, because it can compensate for different biological conditions, converging back to the same outcome.

We show below the implementation in Python for a more modest task – computing how many initial network states converge to a *given* steady state.

```
1   import itertools
2
3   def steady_state_size(G, U, steady_state):
4       ''' % of initial states combinations end up in the attractor '''
5       cnt = 0
6       n = len(G)
7       possible_init_states = [i for i in range(U+1)]
8       for init in itertools.product(possible_init_states, repeat=n):
9           init = list(init) # convert tuple->list
10          trajectory = run(G, U, init)
11          if trajectory[-1] == steady_state: # steady state = attractor
12              cnt += 1
13      percent = 100 * cnt/(len(possible_init_states)**n)
14      return percent
```

Code explained

First we import the library `itertools`. This useful Python library enables us to easily generate all the combinations of states over a given set of possible states. We use this library in line 8, when we call the function `product`. In lines 6 and 7, two parameters are defined. The first one is `n`, the number of nodes in the network. The second is the list `possible_init_states` containing the possible initial states for nodes. This is simply the list `[0,1,2,...,U]`. The function returns all the combinations of length `n` over the possible states, each one as a tuple (of length `n`).

The loop that starts in line 8 takes each such combination (the variable `init`), converts it into a list, and calls the function `run` with `init` as the initial network state. Then it checks if the last step of the simulation (`trajectory[-1]`, as in Python we can access the last element of a list using the index −1) is identical to the steady state we are examining. If so, we increase the counter by 1.

Finally, the function computes the percentage of such combinations that reached our given steady state.

7.4 Biological Case Studies

To end this chapter, we briefly present two biological studies that used the discrete model presented in this chapter. These case studies are based on research published in the two papers included in the reference list.

7.4.1 The Yeast Cell Cycle

In their paper, Li et al. (2004) used a simplified version of the model described here to study key regulators of the cell cycle pathway in the budding yeast *Saccharomyces cerevisiae*. By simplified, we mean Boolean: nodes are limited to just two levels of activity – 0 ("OFF") and 1 ("ON"). In terms of the notations presented earlier, they used $U = 1$. In such a simplified setting, intermediate levels are not expressible.

Their model was used to simulate a single instance of the cell cycle, initiated by an increase in the cell size, after which the cell arrests in the stationary G1 stage, waiting for additional signals. The researchers concluded that the yeast cell cycle network is robust; namely, changes in the state of the system, at the onset of the simulation, do not alter its fundamental behavior. They do so by examining all possible initial states of an 11-node network ($2^{11} = 2048$ possible initial states in total), and show that 1764 of these (86%) led to the same steady state – the main attractor of the system, which resembles a stationary G1 state (see Figure 7.6).

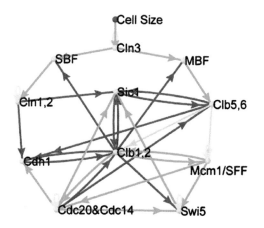

Figure 7.6. The yeast cell cycle network as suggested by Li et al. (2004)

Below is a simulation reproducing this research results:

```
1   nodes = \
2   ['Cln3','MBF','SBF','Cln1,2','Cdh1','Swi5','Cdc20/Cdc14','Clb5,6','Sic1',
3   'Clb1,2','Mcm1/SFF']
4
5   G = [
6       [-1, 1, 1, 0, 0, 0, 0, 0, 0, 0, 0],
7       [ 0, 0, 0, 0, 0, 0, 0, 1, 0, 0, 0],
8       [ 0, 0, 0, 1, 0, 0, 0, 0, 0, 0, 0],
9       [ 0, 0, 0, -1, -1, 0, 0, 0, -1, 0, 0],
10      [ 0, 0, 0, 0, 0, 0, 0, 0, 0, -1, 0],
11      [ 0, 0, 0, 0, 0, -1, 0, 0, 1, 0, 0],
12      [ 0, 0, 0, 0, 1, 1, -1, -1, 1, -1, 0],
13      [ 0, 0, 0, 0, -1, 0, 0, 0, -1, 1, 1],
14      [ 0, 0, 0, 0, 0, 0, 0, -1, 0, -1, 0],
15      [ 0, -1, -1, 0, -1, -1, 1, 0, -1, 0, 1],
16      [ 0, 0, 0, 0, 0, 1, 1, 0, 0, 1, -1]
17      ]
18
19  attractor = [ 0, 0, 0, 0, 1, 0, 0, 0, 1, 0, 0]
20  print(steady_state_size(G, 1, attractor), "% initial vectors end at",
21  attractor)
```

The output of this program is:

```
86.1328125 % initial vectors end at [0, 0, 0, 0, 1, 0, 0, 0, 1, 0, 0]
```

In a later work, Rubinstein et al. (2007) examine this conclusion under a wider range of initial and intermediate states, thus possibly strengthening or weakening the above-mentioned conclusions. $U = 9$ was used, that is, a global upper bound for all states of the nodes. Since an exhaustive exploration of all initial network states is too lengthy (there are 10^{11} such states), initial states were randomly sampled. The attractors reached in this simulation were all steady states (no loops); 86.6% of the simulations converged to one main attractor, which resembles stationary G1 conditions, in agreement with the simulations of the Boolean case. This gives additional support for the robustness of the yeast cell-cycle network as suggested by Li et al., as the system exhibits significant tolerance even to extreme "noise" in the levels of its components.

7.4.2 The Circadian Clock

The majority of organisms in nature have adapted to the rhythmic daily changes in the environment, caused by the earth's rotation around the sun. These mechanisms, termed circadian clocks, are evolutionary conserved and well-studied in a wide variety of organisms. Rubinstein et al. (2016) examined a network representing the main regulators of the vertebrate circadian clock molecular mechanism (see Figure 7.7). The network consists of 22 nodes, which represent a simplified version of the circadian clock components'

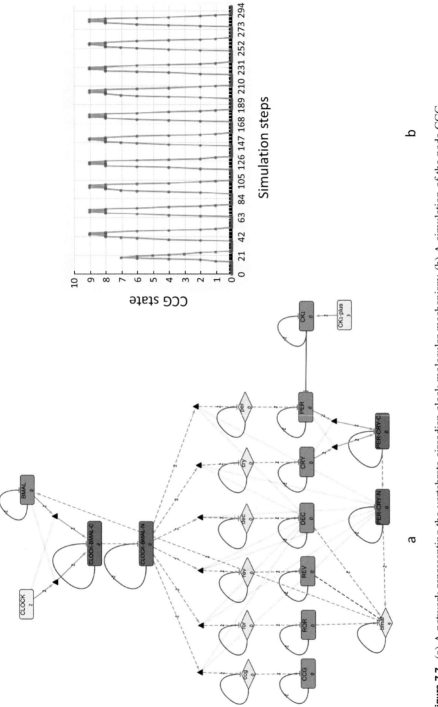

Figure 7.7. (a) A network representing the vertebrate circadian clock molecular mechanism; (b) A simulation of the node CCG

interactions, based on known literature. The simulation of this network converged to a cycle (Figure 7.7(b)), in which nodes states are increased periodically, as expected.

Furthermore, the research examined the influence of sunlight on this system, which is not yet fully understood. At the behavioral level, there exists a well-known experimental phenomenon called the "light phase response curve." It depicts the advance or delay of the circadian clock's phase, resulting from short light pulses introduced during various times of the day. Light pulses at the beginning of the night cause a delay in the circadian rhythm, while pulses occurring later in the night and toward dawn result in a phase advance (see Figure 7.8(a)).

The research explored numerous hypothetical networks reflecting the exact manner by which sunlight affects the system components. This was done by adding a node representing sunlight to the network, connecting it in various ways to the rest of the network, and for each option generating a "computational phase response curve," by activating the sunlight node to "emit short light pulses." Among these hypothetical networks, only very few yielded a curve that resembled the experimental one, such as the one in Figure 7.8(a).

To conclude, this type of analysis may help researchers raise hypotheses and design new experiments to validate them. Further usage of this network of the molecular circadian

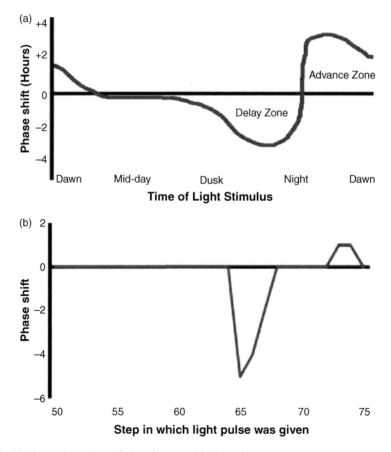

Figure 7.8. (a) Phase advance, graph based on empirical results
(b) A graph based on simulation results that fit the empirical results

clock can take the form of introducing additional mutants and perturbations. For example, knockout of a gene can be simulated by deleting a node from the network or removing its outgoing edges, overexpression of a gene can be simulated by increasing the initial state of the corresponding node, etc.

Reflection

Typically, life scientists working on a specific interaction network are highly familiar with the underlying details of the system and its components. Yet, the complete patterns of interactions between the various components may be too complex to be analyzed manually, and often many facets of the system's behavior remain unknown. Mathematical models, sometimes implemented as simulation tools, provide an important approach in bridging those gaps.

The starting point for using such models and tools is the construction of a network, according to the biological knowledge and understanding of the researcher. The model then enables extensive in-silico simulations – on a scale that is infeasible experimentally – making it possible to tune the system, adding or erasing connections, changing strengths of inter-actions, etc. Such simulations may be used to strengthen or weaken hypotheses regarding the system, and, furthermore, raise new hypotheses and suggest predictions as to the system's behavior under various conditions. Such predictions could later be tested experimentally, followed by modifications in the network, if necessary, according to the accepted scientific paradigm.

There are three main benefits for using models and simulation in biology:

– *Descriptive*: it forces clarity of expression and precision in describing systems/processes/ hypotheses.
– *Analytic*: it promotes understanding of the system and may provide valuable insights.
– *Predictive*: it enables predictions regarding the behavior of the system under various conditions.

Using computational and mathematical models is often an iterative process, as described in Figure 7.9. Biologists gather data from various sources, which serve as a starting point to using an appropriate model and conducting simulations. The results of simulations are compared to known experimental results, which lead to modifications in the model's param-eters. This process repeats as long as the simulation results do not adhere to the experimental ones. When this happens, the model is suitable for making new predictions regarding the biological system. These may lead to new hypotheses, which can be tested in the lab, producing new data to serve as the basis for future simulations.

Models can be roughly divided into two main types: quantitative and qualitative. Quantitative models use the mathematical formalism extensively, and are used to predict specific values that can be compared to actual experimental data, such as, kinetic constants, concentrations, and expression levels. Qualitative models often use a simpler formalism, neglecting some numerical details, and are used to predict trends and more general properties of a system. For example, a qualitative model may be used to test whether, and under which conditions, the system remains in a steady state, or whether a mutation causes a pathway to shut down.

Simulation tools are often used by biologists as "black boxes," with limited famil-iarity of the underlying computational models and their inherent limitations. This may lead to

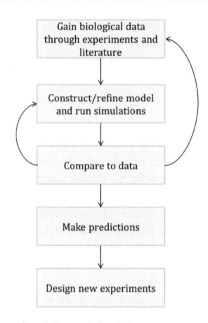

Figure 7.9. The iterative nature of modeling and simulations

misinterpretation of results. While it is infeasible to learn the details of every existing model, we proposed in this chapter a glimpse into the interiors of a specific discrete model and explained its usability to explore regulatory networks.

References and Further Reading

Li, Fangting, et al. "The yeast cell-cycle network is robustly designed." *Proceedings of the National Academy of Sciences* 101.14 (2004): 4781–4786.

Rubinstein, Amir, et al. "Faithful modeling of transient expression and its application to elucidating negative feedback regulation." *Proceedings of the National Academy of Sciences*, 104.15 (2007): 6241–6246.

Challenge Yourself

Problem 1 Detecting simulation cycles

The function `run` you saw in this chapter assumes a simulation converges to a steady state; that is, two consecutive network states which are identical. We now want to enhance this function to detect simulation cycles as well. A simulation is cyclic if it contains two identical (but not necessarily consecutive) network states. For example, see the network from exercise 3. If you start with X,Y = (1,0), after one simulation step you go to X,Y = (0,1), and right after that you reach X,Y = (1,0) again. Clearly, the simulation will flip between (0,1) and (1,0) endlessly.

To detect a cycle, we need to change the condition `new_states != states` in the loop of the function `run`. Instead of checking for two consecutive network states, we need to check if the new network state has already appeared before.

(a) Modify the function `run` accordingly. Make it halt if a steady state *or a loop* has been detected.

(b) Modify accordingly the function `pretty_print`. When a simulation has ended with a cycle, it will print the steps at which the cycle began and ended (the last step).

Execution examples, for the network from Figure 7.4:

```
>>> U = 9
>>> init = [9, 0, 5, 1]

>>> G = [  [ 0, 1, 0, 0],    # node A
           [ 0,-1, 1, 0],    # node B
           [-1, 0,-1, 0],    # node C
           [ 1, 0, 0, 0]  ] # node signal

>>> pretty_print(run(G, 9, init))
step:   network state:
0 [9, 0, 5, 1]
1 [8, 1, 4, 1]
2 [7, 2, 3, 1]
3 [6, 3, 2, 1]
4 [5, 4, 3, 1]
5 [4, 5, 4, 1]
6 [3, 4, 5, 1]
7 [2, 3, 4, 1]
8 [1, 2, 3, 1]
9 [0, 1, 2, 1]
10 [0, 0, 1, 1]
11 [0, 0, 0, 1]
12 [1, 0, 0, 1]
13 [2, 1, 0, 1]
14 [3, 2, 1, 1]
15 [3, 3, 2, 1]
16 [2, 3, 3, 1]
17 [1, 2, 3, 1]
the simulation ended in a cycle between steps 8 to 17
```

Part IV: Images

– "A picture is worth a thousand words."

Images are used extensively nowadays. With the digital revolution, we easily generate, send, and observe images, all at a very low cost. Images are also used extensively in biological research and in the medical clinic. Biological images are studied mainly in basic science research, mostly at the cellular and molecular level. Medical images are used for clinical purposes, and focus on the tissue and organ level. Both kinds of images are often complicated and heterogeneous, and analyzing them requires sophisticated computational techniques. The goal of these techniques is to extract meaningful knowledge about the image content. For example, the automated identification of objects in the image (such as cells, intracellular components, or a cancer tumor), tracking cells, organelles, or cancer cells in consecutive video frames, or phenotype identification by object properties (size, light intensity, shape, etc.). Images also require large volumes of storage, raising the need for efficient compression algorithms, in order to store them.

The bio and medical image explosion we are witnessing is due to several technological advances in recent decades. Nobel Prizes for Chemistry were awarded for two important breakthroughs. The 2008 Nobel in Chemistry was given for the discovery of green fluorescent protein (GFP), which is used as a tagging device:

This year's Nobel Prize in Chemistry rewards the initial discovery of GFP and a series of important developments which have led to its use as a tagging tool in bioscience. By using DNA technology, researchers can now connect GFP to other interesting, but otherwise invisible, proteins. This glowing marker allows them to watch the movements, positions and interactions of the tagged proteins.

(from the Nobel prize website:
http://www.nobelprize.org/nobel_prizes/chemistry/laureates/2008/press.html)

The second technological breakthrough, which was awarded the Nobel Prize in 2014, was the development of the high-resolution optical microscopy:

For a long time optical microscopy was held back by a presumed limitation: that it would never obtain a better resolution than half the wavelength of light. Helped by fluorescent molecules the Nobel Laureates in Chemistry 2014 ingeniously circumvented this limitation. Their ground-breaking work has brought optical microscopy into the nanodimension.

(from the Nobel prize website:
http://www.nobelprize.org/nobel_prizes/chemistry/laureates/2014/press.html)

In addition, high-throughput microscopy systems, together with the decreasing cost of storage devices (memory), enable us to collect and store an overwhelming number of images. This explosion of imaging data makes it necessary to develop efficient computational approaches to analyze and visualize the data. Obviously, automated systems can (sometimes) perform better than humans, as they are superior in tedious, systematic and reproducible analyses, potentially involving numerous parameters and metrics (e.g., counting hundreds of cells of certain shape in an image). Also, they can detect fine details and identify complex patterns that may be hidden to the human eye (or brain). Automated systems are also unbiased (excluding bias introduced by the humans who designed them), as opposed to an observation-based (subjective) analysis by humans, whose focus of attention is directed to events that match their expectations.

However, the development of computational solutions to various biological and medical image problems is far from trivial. As explained in the essay "Computer Vision in Cell Biology" by Gaudenz Danuser:

the key challenge in developing a computer vision system for cell biology is to conceptualize in a programmable framework the parameters, rules, and models required to characterize cell biological function. This requires profound familiarity with the cellular mechanisms, the range of valid hypotheses, and the expected shifts in phenotypes that may distinguish one model from another. It often also requires familiarity with the molecular techniques for labeling and manipulating cells and with the microscopy itself, as the experimental design directly influences whether image data can be analyzed in terms of a particular question.

(Gaudenz Danuser, Computer Vision in Cell Biology, Cell, Volume 147, Issue 5, 23 November 2011, Pages 973–978, doi:10.1016/j.cell.2011.11.001)

It should be emphasized that advances in machine learning, and especially the so-called deep learning, together with stronger processors (GPUs, and now special machine learning processors), coupled with huge numbers of tagged examples, are paving the road to routine and correct analysis of images, which were deemed impossible not long ago.

In this part of the book, we will try "demystifying" the fundamental properties of the digital image world, and understand the basic details of how images are represented and handled by a computer. Yet, we will not get into specific, advanced technological details, or state-of-the-art image analysis tools. Even though many such excellent tools exist, we aim here at the development of computational thinking in the context of images. Therefore, we focus on the basic ideas and concepts, rather than on the tool handling.

Chapter 7 explains what a digital image is, and how it can be represented and stored in a computer. In chapter 8, some basic notions in image processing are discussed. Image processing means, informally, the manipulation and/or analysis of an image.

Image-Related Python Packages

The Python Imaging Library (`PIL`)

The code examples we will show require installing an imaging package. It is highly recommended for readers who wish to follow the implementation details and experiment with the code to install this package first. All installation instructions appear on the website. `PIL` is Python's library for image processing. This is the main package that this part of the book is based on. All the code examples require it. Once `PIL` is installed, the following command should run smoothly:

```
>>> from PIL import Image
```

The Scientific Computing Library (`scipy`)

For chapter 8, the section on labeling, you will need to install Python's scientific library, one of Python's most commonly used libraries. Once `scipy` is installed, the following command should run smoothly:

```
>>> from scipy.ndimage.measurements import label
```

Optional: The `swampy` Package for GUI (Graphical User Interface)

In order to ease learning and exploration for the readers, we have built a simple Python-based tool with graphical interface (see Figure IV.1). A graphical interface will enable the reader to "play" with the various functions presented in this part of the book more fluently, by clicking buttons and defining parameters in text boxes, rather than running code directly. Such an interface is termed **GUI – graphical user interface**. Note that everything doable with this tool is also doable via executing Python code directly. Indeed, all the code and execution examples presented here use the regular format of this book, so readers may choose not to use the graphical user interface option.

The implementation of the GUI tool consists of all the functions that we will meet in the following two chapters, in addition to the code responsible for the interface itself (the tool's layout, buttons, text boxes, etc.). Using the GUI tool does not require understanding its implementation in Python. All you need is to install a Python package called `swampy`

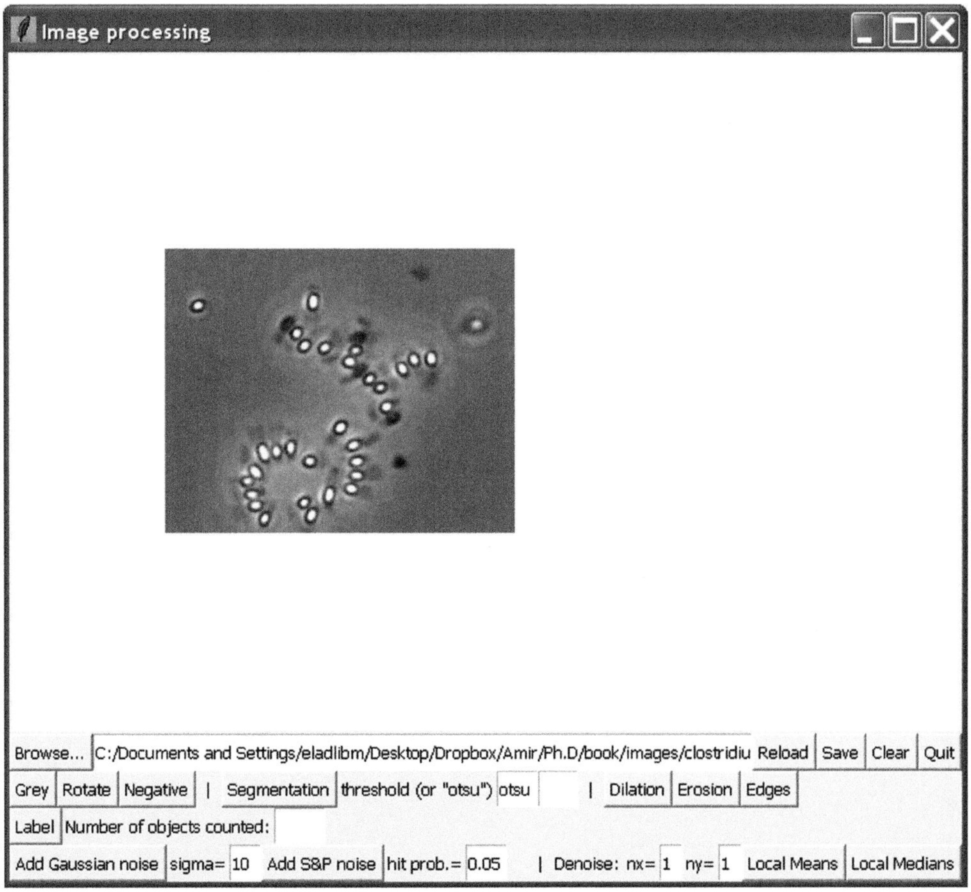

Figure IV.1. A graphical user interface (GUI) for this part of the book

(see instructions on our website). *Swampy* is a collection of Python programs, designed for educational purposes, which provides a simplified interface to the functions and classes in `Tkinter`, a popular Python package for writing GUI-based programs.

After installing `swampy`, all you need to do to follow this chapter's material is to download the file Gui.zip (also from the website), extract it to some location on your computer, and run the file Gui.py.

8   Digital Images Representation

Computers allow us to store an enormous amount of data. Data come in several types, such as numbers, sound, and images. But how are these types of data represented in the computer's memory?

The computer stores bits (binary digits, zeros, and ones), in billions of tiny electrical or magnetic devices. Bits can represent integers. For example, the sequence 10100 in binary represents the decimal number 20, and 10101 represents 21. Other types of data are actually represented using integers. For example, there is a table called Unicode, which maps characters to integers, and this is how the computer represents text.

In this chapter, we will explain how colors and light intensities – that is, images – are represented using integers. We will not get into the technological issues of, for example, how color is presented on your screen, but merely focus on the logical level – how to store visual data for analysis and processing.

8.1 What Is a Digital Image?

In the physical world, any quantity measurable through a sequence in time or space can be viewed as a signal. Examples include radio, telephone, radar, sound, images, video, sensor data, and other sources. Signal processing is the application of mathematical techniques for the extraction, transformation, and interpretation of signals. Generally, signal processing may take two major flavors: discrete (digital) or continuous (analog).

One important branch of signal processing deals with images, such as digital photos, microscopic images, medical images, and astronomical images. Even images of light frequencies that are unseen to the bare human eye exist and are being used, especially in scientific research (e.g., x-ray telescopy).

So what is a **digital image**? It is a discrete, numeric representation of a "real" image: the image is divided into a specified number of cells, called **pixels** (picture elements). The numbers of rows and columns are usually denoted by n and m, respectively. So we can say that a digital image is encoded (represented) by an n x m matrix M, where the pixel at position x,y for some $0 \leq x \leq n$-1 and $0 \leq y \leq m$-1 conveys the numeric information about the light intensity and color components of that location in the image (see Figure 8.1).

For **color images** in RGB (Red–Green–Blue) format, each pixel specifies how much "red," "green," and "blue" that position in the image contains (see Figure 8.2). There are other color representation formats, which we will not discuss here.

By way of contrast, in **gray-level images** pixels correspond only to the light intensity at that image location. So each pixel contains a single value, specifying how dark/light the pixel is (instead of three values in RGB images).

For simplicity, in this chapter we will deal only with gray-level images. The standard setting is 256 gray levels for each pixel, but there are settings with more gray levels or less (see Figure 8.3 for an example).

An important related notion is **pixel resolution**. Resolution is the capability of the sensor (camera lens, human eye) to observe or measure the smallest object clearly with

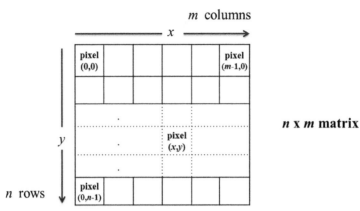

Figure 8.1. A digital representation of an image

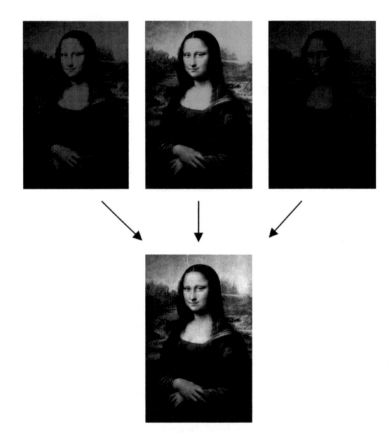

Figure 8.2. RGB components of an image (http://www.csse.uwa.edu.au/~wongt/matlab.html)

distinct boundaries. Resolution depends upon the physical size of a pixel: for the same area covered by an image, higher resolution means lower pixel size. For example, in Figure 8.4 the resolution is decreased as we move from right to left, from a 100×100 resolution (each row and each column consists of 100 pixels, $100^2 = 10{,}000$ pixels altogether) to 50×50, 20×20, and so on. In the 50×50 image for example, each pixel is twice the height and width of a

Figure 8.3. 256 gray-level image

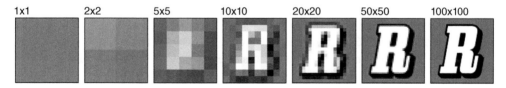

Figure 8.4. Pixel resolution (image from Wikipedia)

pixel in the 100×100 image, and the resolution is half. In the extreme case of the 1×1 image, it consists of a single pixel. The higher the resolution is, the better we can observe fine details of the shapes in the image (see Figure 8.4).

The term **bit depth** relates to the number of bits used to represent each pixel. For example, in a standard gray-level image with 256 different gray levels (see Figure 8.5(a)), each pixel will have eight bits (with eight bits, one can represent $2^8 = 256$ different values). If we use only 1 bit per pixel (that is, the bit depth is 1), then each pixel can assume only two different values, and thus the image becomes a black and white image (for example, 0 represents black, 1 represents white) (see Figure 8.5(b)). In general, with k bits per pixel, we can represent 2^k gray levels for each pixel. In some cases, such as in medical images, we may require a higher bit depth, in order to distinguish, for diagnostic purposes, between very similar light intensities in the image. For example, images produced by x-ray detectors, such as CT images, typically contain between 12 and 16 bits per pixel (which corresponds to

(a)

(b)

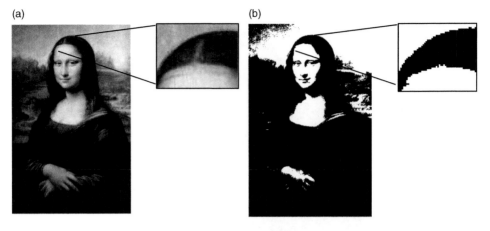

Figure 8.5. (a) 256 gray levels (0–255) (b) Two gray levels (0–1)

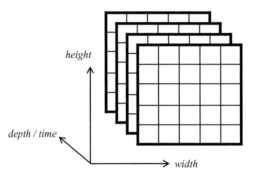

Figure 8.6. 3D image

between 4,096 and 65,536 levels of gray). However, it is claimed that a human observer can discriminate between at most a few hundred shades of gray in optimal conditions (some estimations are lower, depending also on the background, distance from the image, etc.). Therefore, bit depths above 8 are usually intended for computerized analyses rather than mere human inspection.

So far, we have referred to an image as a 2-dimensional (planar) entity. However, images of higher dimensions are very common as well (see Figure 8.6). For example, a video is merely a sequence of 2D images, called frames, taken over time (thus the third dimension is time). In biological and medical images, it is common to use various labels, such as fluorescent probes, as additional dimensions. For example, two images capturing the same object at the same time, but with green and blue fluorescent markers can be thought of as 3D images, where the third dimension is the marker color.

Digital images with high pixel resolution and bit depth take up lots of computer memory. For example, a 1024×1024 gray-level image with a standard 8 bit depth takes $1024 \cdot 1024 \cdot 8$ bits. This is $2^{10} \cdot 2^{10} \cdot 2^3 = 2^{23}$ bits, or $2^{10} \cdot 2^{10} = 2^{20}$ bytes $= 1\text{MB}$ (a byte is eight bits). Given the extreme ease and popularity of digital photos in recent years, it is no wonder that images are a serious threat to the limit of digital storage devices. This motivates the need for **compressing** images.

> ### Exercise 1
>
> A one minute long video is made out of a sequence of 2D images captured every second. Each image is of size 512x512 pixels and depth 12 bit/pixel.
> How much memory does this video consume (without any compression)?

Generally speaking, compression takes an object A and converts it to an object B, which consumes less memory but still contains the meaningful information in the original object. In this process, some of the information in A may be lost, in which case the compression is termed **lossy**. Otherwise, we call it **lossless**. The various image formats you are probably familiar with, such as jpg, tiff, png, bmp, gif, differ by the type of compression applied to the original image. The bmp format is lossless, while the other formats are lossy (tiff can be both, depending on some parameter settings).

Most people will not be happy to use a compression method that simply erases a portion of the image to reduce its size. Instead, less crude compression methods are employed, where the loss of information is hardly visible. For example, the jpg format partitions an image into squares of 8-by-8 pixels. Most such squares will exhibit only gradual, moderate changes between neighboring pixels, especially in smooth areas of the image (see Figure 8.7). These gradual changes can be well approximated by far fewer bits than required for $8 \cdot 8 = 64$ pixels. Furthermore, typically the reduced amount of information is not perceived as seriously degrading the image quality, while a factor of 10 (or even more) saving in space can be achieved.

8.2 Getting Started with `PIL`

Let's begin right away with some code.

```
>>> from PIL import Image
>>> im = Image.open("./some_image.jpg")

>>> im.size       <= get image dimensions
(388, 541)        <= width, height

>>> im.show()     <= display

>>> rot = im.rotate(45)    <= rotate

>>> rot.show()

>>> im.save("./my_image", "jpg")    <= save

>>> region = im.crop((100,300,250,400))    <= crop the image as in figure 8.8

>>> region.show()
```

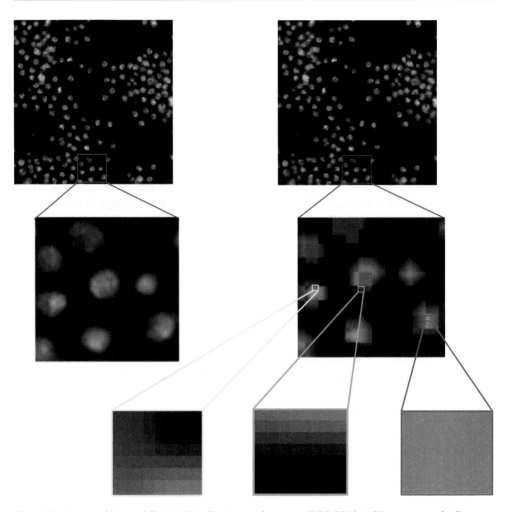

Figure 8.7. Image of Drosophila Kc167 cells (source: image set BBBC002v1 [Carpenter et al., Genome Biology, 2006] from the Broad Bioimage Benchmark Collection [Ljosa et al., Nature Methods, 2012]), with permission from David Root/Anne Carpenter and under a Creative Commons license.

Left: Original image. Right: Highly compressed image in jpeg format. Three 8×8 pixel squares are shown: In the blue square, all pixels are almost identical. In the green square, pixels mostly change from top to bottom. In the yellow square, pixels change in both directions.

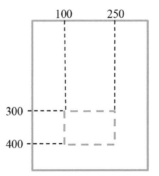

Figure 8.8. Cropping an image in PIL

Code explained

We first imported `PIL` into the workspace, or more specifically part of the `PIL` package called `Image`. For this to work, we had to install `PIL` first, as explained earlier. All the functions used above belong to the `Image` package.

`Image.open` will open an existing image file, in one of the common formats, such as jpg, bmp, etc. The function takes the location of the image in the file system, and `./` (dot slash) is used to access the current directory. Then, the variable `im` is an image type object as defined by the `Image` package.

An image type object has various fields and functions. For example, one field is called `size`, and it stores a tuple with the image's widths and height. Several functions are exemplified above, such as `show`, `rotate`, `save`, and `crop`. The command `im.show()` will present the image. Try the other functions and observe the result.

As mentioned, we will be working in this chapter with gray-level images. Each pixel will hold one of 256 gray levels, so we will need eight bits per pixel. It is common to let the value 0 denote minimal light intensity and 255 maximal one. Thus, for gray-level images, 0 will denote complete black, while 255 complete white. For an example see Figure 8.9.

We can use the function `convert` to convert any image (color, gray, black, and white), into a 256 gray-level image:

```
>>> im = im.convert('L')   <= convert to 256 gray levels (0-255)
>>> im.show()
```

(a)

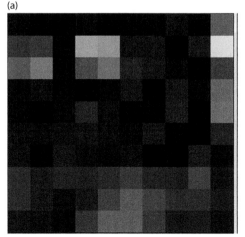

(b)
```
38,  26,  21, 36,  19,  28,  33, 44, 31, 112,
77,  83,  34, 168, 159, 48,  50, 14, 55, 211,
112, 137, 34, 101, 129, 62,  54, 40, 21, 86,
41,  46,  35, 19,  35,  52,  18, 57, 39, 123,
38,  16,  38, 67,  45,  21,  29, 59, 10, 130,
45,  43,  46, 51,  44,  39,  53, 31, 24, 64,
47,  30,  54, 45,  40,  46,  23, 26, 58, 40,
71,  57,  66, 63,  70,  84,  65, 62, 91, 49,
72,  55,  43, 57,  90,  111, 92, 73, 74, 56,
47,  45,  36, 78,  114, 113, 81, 54, 57, 44
```

Figure 8.9. (a) 10*10 pixels of 256 gray-level image (b) The corresponding pixel values

> **Exercise 2**
>
> Open a color image of your choice using `Image.open()`, then convert the image into a 256 gray-level image, and observe the result using the function `show`.

You may wonder how this conversion is actually performed. Without going into all the details, if the image is a color image, Python computes a weighted average of the three RGB components (each color component is given a different weight).

> **Exercise 3**
>
> In black and white images, black is represented by 0 and white by 1. How do you think Python converts such images into 256 gray-level images?

In order to analyze and manipulate an image, we need access to its pixels. To that end, we can `load` the matrix that represents the image:

```
>>> im = Image.open("./my_image.jpg").convert('L')
>>> mat = im.load()
>>> mat[0,0]  <= upper left corner
31
>>> mat[0,0] = 255
>>> mat[0,0]
255
>>> mat[0,0] = mat[0,1] = mat[1,0] = mat[1,1] = 0 <=…
... a small white square at the upper left corner
>>> im.show()  <= im was automatically affected by changes to mat
```

> **Code explained**
>
> The variable `mat` is the matrix that represents the image in the variable `im`. We can now access each pixel using two indices for rows and columns. For example, the upper leftmost pixel (the pixel at position 0,0) stores the value 31, which is dark gray (but not completely black). Note that the indices are provided in square brackets $[i,j]$, where $0 \le i < height$ and $0 \le j < width$. We can also change the value of pixels, as shown above. When we change the pixels in the matrix, the corresponding image is automatically affected.

8.3 Generating Synthetic Images

We will now see several examples for synthetically generated images, some of which are rather surprising. Each example will include a single function that generates a specific pattern.

First, this is how we create new images:

```
>>> im = Image.new(mode='L', size=(100,50), color=255)
```

Code explained

The function new of the Image package returns a new image. It takes three parameters: The first one is the "mode," or format of the image, where L means 256 gray-level image (there are other options, such as 1 for a black and white image of bit depth 1). The second parameter is the image size as a tuple (width, height). The third one tells how to initialize the pixels in the new image. For example, color=255 will create an image with all pixels white.

Now, observe the following function:

```
1   def f1(w,h):
2       ver_lines = Image.new(mode='L', size=(w,h), color=255)
3       mat = ver_lines.load()
4
5       for x in range(w):
6           if x % 10 == 0:
7               for y in range(h):
8                   mat[x,y] = 0
9
10      return ver_lines
```

Code explained

In line 2, we created a new white image, of width and height as specified in the parameters of the function. We called the new image ver_lines, for a reason to be clear soon. In line 3, we loaded the representing matrix so we can change the pixels and create some patterns. Line 5 goes over all x values between 0 and w−1, that is, x represents indices of columns in the image. For each column index, line 6 checks if it's a multiplication of 10, that is 0,10,20, and so on. If it is, then line 7 goes over all the values of y from 0 to h−1, that is y represents an index of a row. Line 8 colors the pixel x,y black. Finally, the function returns the newly generated image.

We can now check the resulting pattern, for example:

```
>>> img = f1(100,50)
>>> img.show()
```

We obtain the image in Figure 8.10, which consists of 10 vertical lines at columns 0,10,20,...,90:

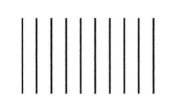

Figure 8.10. Output for `f1(100,50).show()`

Figure 8.11. A black diagonal on white background

Exercise 4

Inspect the following function f2. Try to understand which image it creates, then check it, for example by calling `f2(100).show()`.

```
def f2(n):
    surprise = Image.new(mode='L', size=(n,n), color=255)
    mat = surprise.load()

    for x in range(n):
        for y in range(n):
            mat[x,y] = x%256

    return surprise
```

Have you ever wanted to draw a black diagonal on a white square? (See Figure 8.11.) The trick here is to note that on the main diagonal (going from upper left to lower right), the index of the row equals the index of the column. That is, if x == y, then the pixel x, y must be located on the main diagonal.

```
1   def diagonal1(n):
2       diag = Image.new(mode='L', size=(n,n), color=255)
3       mat = diag.load()
4
5       for x in range(n):
6           for y in range(n):
7               if x-y == 0: # x == y
8                   mat[x,y] = 0
9
10      return diag
```

Actually, we could write this function more efficiently. Instead of going over all the cells of the matrix (n^2 for a matrix with n rows and n columns), we could do it with only a single loop, directly accessing the pixels on the diagonal:

```
1  def diagonal2(n):
2      diag = Image.new(mode='L', size=(n,n), color=255)
3      mat = surprise.load()
4      for x in range(n):
5          mat[x,x] = 0
6      return diag
```

Code explained

Instead of going over all pixels, and asking for each one "are you on the diagonal?" (in Python: `if x == y`), what we did here was to go over all values of x from 0 to `n-1`, and color the pixel at row x and column x black.

Note that while the first solution has to perform n^2 iterations in total, the second one only performs n, and thus is much more efficient.

Exercise 5

(A) Suppose we changed line 7 in diagonal 1 from

```
if x-y == 0:
```

to

```
if x-y == 10:
```

What would the difference be?

(B) The same for

```
if x-y <= 10:
```

(C) The same for

```
if -10 < x-y < 10:
```

Exercise 6

Observe the following function. Can you explain the resulting image (shown in Figure 8.12)?

```
def diagonals(n):
    surprise = Image.new(mode='L', size=(n,n), color=255)
    mat = surprise.load()

    for x in range(n):
        for y in range(n):
            mat[x,y] = (x-y) % 256

    return surprise
```

Figure 8.12. Result of `diagonals(512).show()`

The next two examples are rather surprising, and really cool. Note that it is rather hard to predict the resulting patterns in these cases. Although the functions are not complicated, and it is easy to compute the gray level at any position in the image, viewing the whole image in mind is difficult (at least for the authors …) (see Figure 8.13 for the image generated by `circles` and Figure 8.14 for the image generated by `product`).

```
1 def circles(n):
2     surprise = Image.new(mode='L', size=(n,n), color=255)
3     mat = surprise.load()
4
5     for x in range(n):
6         for y in range(n):
7             mat[x,y] = (x**2 + y**2) % 256
8
9     return surprise
```

```
1   def product(n):
2       surprise = Image.new(mode='L', size=(n,n), color=255)
3       mat = surprise.load()
4
5       for x in range(n):
6           for y in range(n):
7               mat[x,y] = (x*y) % 256
8
10      return surprise
```

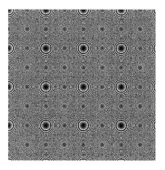

Figure 8.13. Output for `circles(500).show()`

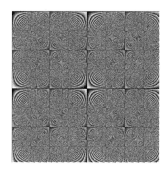

Figure 8.14. Output for `product(500).show()`

8.4 Simple Image Manipulations

In the last section of this chapter, we show some simple functions that manipulate an existing image (in contrast to generating a new image as in the previous section). All these functions get as a parameter a 256 gray-level image.

```
1 def add(im, k):
2       w,h = im.size
3       mat = im.load()
4       new = im.copy()
5       mat_new = new.load()
6
7       for x in range(w):
8           for y in range(h):
9               mat_new[x,y] = (mat[x,y] +k) % 256
10
11      return new
```

Code explained

The function receives an image `im`, and an additional parameter `k`, which is an integer. In lines 4 and 5, it creates a copy of the input image and loads the underlying matrix.

In the nested loop in lines 7 and 9, the function goes over all the positions in the images. It stores in the new image at position `x,y` the value of the original image at that position plus `k`. To keep all pixel values between 0 and 255, it also uses `%256` (the remainder of 256 can be a number between 0 and 255).

This function simply makes pixels whiter by `k`, except for pixels whose original value was already very light: pixels that are at most `256-k` become dark. For example, if the original pixel at some position was 250 (almost completely white), and `k = 10`, the new pixel will become (250+10)%256 = 260%256 = 4 (almost completely black).

You can see an example in Figure 8.15 (note how the originally very light pixels on the snow at the base of the mountain become dark).

Here is a function that creates a negative of an image (see the result of applying it on that same mountain picture, in Figure 8.16):

```
1 def negate(im):
2       w,h = im.size
3       new = im.copy()
4       mat = im.load()
5       mat_new = new.load()
6
7       for x in range(w):
8           for y in range(h):
9               mat_new[x,y] = 255 - mat[x,y]
10
11      return new
```

(a)

(b)

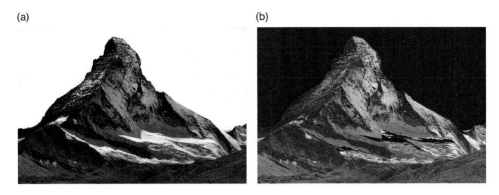

Figure 8.15. (a) Original image (b) Output for `add(im, 30).show()` (Credit: Andrew Merry / Moment / Getty Images).

(a)

(b)

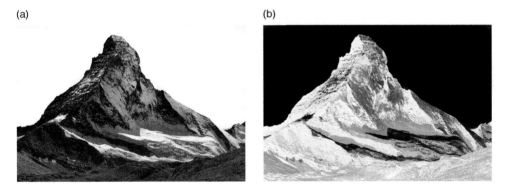

Figure 8.16. (a) Original image (b) Output for `negate(im).show()`

> ### Code explained
>
> If we take a pixel of value $0 \leq p \leq 255$, then $255 - p$ is the negative of it. For example, 0 becomes 255, 1 becomes 254, and 255 becomes 0.

Finally, the following function rotates an image upside down:

```
1   def upside_down(im):
2       w,h = im.size
3       new = im.copy()
4       mat = im.load()
5       mat_new = new.load()
6
7       for x in range(w):
8           for y in range(h):
9               mat_new[x,y] = mat[x,h-y-1]
10
11      return new
```

    Code explained

Suppose the input image is of size (50,100): width = 50 and height = 100. To turn the image upside down, we need the first row to become the last, the second to become the second to last, and so on. Suppose we have an image of height = 100. Then row 0 should become row 99, row 1 should become row 98, etc. More generally, for an image of height h we need row y to become row h-y-1. Therefore, in line 9 the pixel x,y in the new image is copied from the pixel x,h-y-1 in the original image. Note that the x-coordinates remain unchanged, only the y-coordinates change.

```
>>> im = Image.open("Narcissus.jpg")
>>> im_flipped = upside_down(im)
>>> im_flipped.show()
```

Examine the original image and the image after the above commands applied on it in Figure 8.17.

    Exercise 6

Write a function `rightside_left` that switches an image horizontally (right becomes left and vice versa).

Exercise 7

What would change in the output of `upside_down`, if we worked on the input image itself, rather than on a copy of it? That is, if we changed line 9 to:

```
mat[x,y] = mat[x,h-y-1]
```

(a)   (b)

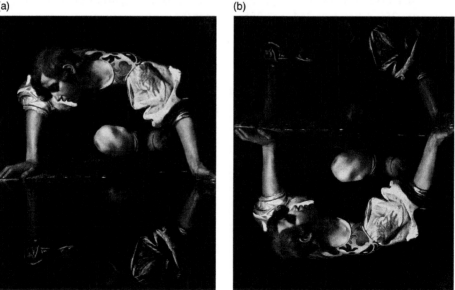

Figure.8.17. (a) Narcissus by Caravagio (b) Output `im_flipped.show()`

Reflection

As we saw, in the end a digital image is simply a matrix of integers. If we ignore the electro-optical details of how colors are presented on a physical screen (which are the topic for a different book), things are rather simple. We use a sequence of bits to represent different values in the computer's memory. These values can be numbers, text, colors, and other types of data. When the data are visual, each value corresponds to some color or grayscale level. Each position in the image has its value, and together they form an image!

In this chapter, we learned how to create "synthetic" images. The idea was simple: initialize an empty (white) image of the desired width and height, and then go over the image's pixels and assign a value to each pixel according to some rule. We also learned how simple image manipulation can be done, by changing pixels methodically.

Challenge Yourself

Problem 1 Playing with synthetic images

Write functions that generate the images in Figure 8.18, each a 9 × 9 pixels image. The function should look very similar to the images from Section 8.3 in their structure.

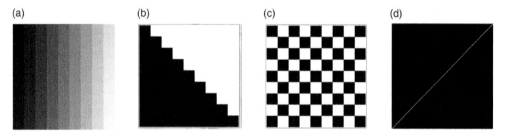

(a)   (b)   (c)   (d)

Figure 8.18. Synthetic images for problem 1

9 Image Processing

The term image processing refers to the manipulation or analysis of digital images. While manipulation alters the image in some way to make it more meaningful or informative, analysis merely extracts information from the image without changing its pixels. This chapter describes some very basic notions in image processing, namely segmentation, morphological operators such as erosion and dilation and noise reduction, and labeling. Image processing is really a whole field of expertise, involving rather sophisticated mathematical methods. This chapter aims to familiarize the reader with the field, setting the ground for further exploration of this fascinating topic.

The main biological example used in this chapter is that of cell quantification – counting the number of cells in a microscopy image. This is one of the most common tasks that biologists who work with images are faced with as part of their daily research routine. Obviously, merely counting is not always enough, and one is often required to differentiate between several types of cells, tracking their movements, etc. Things are often complicated, because images come in different types, resolutions, bio-labeling techniques, quality, etc. Therefore, there is no such thing as "the best algorithm" for a specific task, and many times an algorithm is tailored for a specific type of images.

9.1 Segmentation

The first and perhaps the most basic image-processing technique is segmentation. Image segmentation is the process of partitioning a digital image into segments – sets of pixels (also known as superpixels), that share similar colors or other properties.

The goal of segmentation is to simplify and/or change the representation of an image into something that is easier to analyze. Image segmentation is critical for many subsequent processes, such as object recognition, shape analysis, and tracking objects. It is typically used to locate objects and boundaries (lines, curves, etc.). In medical images, segmentation facilitates locating tumors and other pathologies, measuring tissue volumes, and diagnosis of anatomical structure.

9.1.1 Binary Segmentation by Thresholding

Suppose the image contains two classes of pixels denoted **foreground** and **background**, and these two classes have distinct, different light intensities: the background is much darker than the foreground (see Figure 9.1(a)). Turning a gray-scale image into a binary image (black and white) is called binary segmentation (see Figure 9.1(b)). One way of doing this is to apply a threshold to all pixels: every pixel above the threshold becomes white (255), whereas the other pixels become black (0). Generally, one can apply more than one threshold, creating more than two types of segments.

(a)   (b)

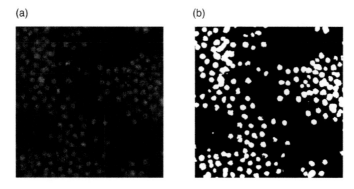

Figure 9.1. (a) Drosophila Kc167 cells, https://data.broadinstitute.org/bbbc/image_sets.html
(b) Binary segmentation, threshold = 27

```
1   def segment(im, thrd):
2       ''' Binary segmentation of image im by threshold thrd '''
3       width, height = im.size
4       out = Image.new(mode='L',size=(width, height))
5       mat = im.load()
6       out_mat = out.load()
7
8       for x in range(width):
9           for y in range(height):
10              if mat[x,y] >= thrd:
11                  out_mat[x,y] = 255 # white
12              else:
13                  out_mat[x,y] = 0 # black
14
15      return out
```

 Code explained

This function takes an image `im`, and a threshold `thrd` by which the image will be segmented. In line 3, we extract the dimensions of the input image. Then, in line 4, we initialize a new gray-scale image with the same dimensions. Lines 5 and 6 extract the matrices representing the input and output images. Lines 8–13 apply the segmentation. We go over all the pixels in the input image, and for each one – if it is at least as light as the threshold – we assign 255 (white) to the corresponding pixel in the output image. Otherwise, we assign 0 (black).

Exercise 1

Modify the function `segment(im)` to perform a ternary segmentation, instead of a binary. In the new function, `segment3` will get two thresholds *thrd1* and *thrd2*. A pixel whose value is *p* will be changed in this way:

- If $p <$ *thrd1*, the pixel will become black (0)
- If *thrd1*$\leq p <$*thrd2*, the pixel will become gray (128)
- Otherwise, *thrd2*$\leq p$, and the pixel will become white (255).

The key issue in thresholding is to select the appropriate threshold, such that the objects in the image will be preserved and intensified. For example, in Figure 9.2 you can examine the results of segmentation with various threshold values on Figure 9.1(a). You may notice that when the threshold is too low (12 in this case), areas in the image where cells are densely populated become "bulbs". In contrast, when it is too high (42) some cells are lost (those whose brightness was low in the original image).

Now try it yourself! You can use a code of the following structure:

```
im = Image.open("...").convert('L')

for th in [20,40,60]: <= Binary segmentation with various thresholds
    out_im = segment(im, th)
    out_im.show()
```

9.1.2 Otsu's Method

A good threshold for segmentation is one that:

- Minimizes differences within each segment.
- Maximizes differences between segments.

One method to find a good threshold is called Otsu's method. This method assumes that the image contains two classes of pixels (foreground pixels and background pixels), and calculates the optimum threshold separating the two classes. The first stage is to calculate the histogram of gray-level values (i.e., the number of pixels for every value of gray). For example, examine Figure 9.3.

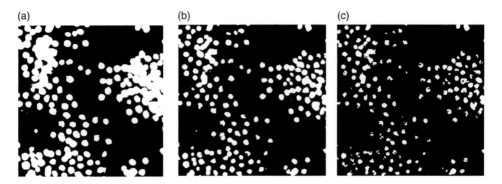

(a)   (b)   (c)

Figure 9.2. (a) Threshold = 12 (b) Threshold = 27 (c) Threshold = 42

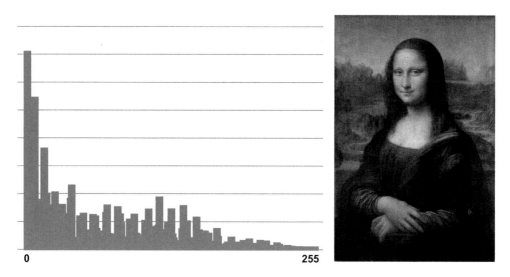

Figure 9.3. A gray-level histogram of 256 gray-level image. The Y-axis represents the number of pixels for each gray-level value

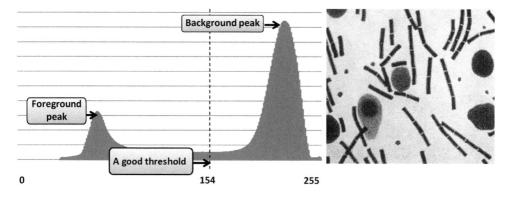

Figure 9.4. Image histogram for a photomicrographic view of Bacillus anthracis bacteria. Foreground and background are easily separable.
Credit: Smith Collection / Gado / Getty Images.

Otsu's method relies on the assumption that the foreground and the background of the image differ substantially in their brightness. This assumption is not true in many cases, as in the example in Figure 9.3 (the foreground in the image contains both dark and bright pixels, and so does the background of the image). However, when this assumption holds, there are expected to be two peaks in the gray values of an image histogram. One peak corresponds to the background and the other to the foreground or the objects in the image. Such image histograms are called **bi-modal**. In this case, the lowest mid-point between these two peaks would be a good choice for a threshold. For example, examine Figure 9.4.

When the difference between foreground and background is less sharp, the peaks may be partly overlapping, as you can see in Figure 9.5. However, in some cases, there will be no such two peaks at all (in which case Otsu's method will be inapplicable), as you can see in Figure 9.6.

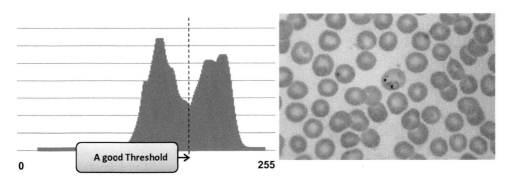

Figure 9.5. Image histogram for RBC with Two P. Vivax Rings, Photomicrograph of a double infected red blood cell with the malaria parasite Plasmodium vivax. Foreground and background are still separable.
Credit: Smith Collection / Gado / Getty Images

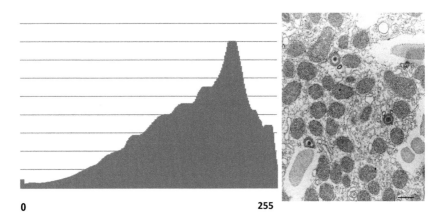

Figure 9.6. Transmission electron microscopy of Protozoa, with no visually separable peaks in the histogram.
Credit: Richard Allen (University of Hawaii)

PIL has a function called `histogram` that returns an image histogram (simply write `im.histogram()`). But even if it didn't, we could write one of our own:

```
1   def histogram(im):
2       ''' Return a histogram as a list,
3           where index i holds the number of pixels with value i '''
4       mat = im.load()
5       width, height = im.size
6       hist = [0] * 256
7
8       for x in range(width):
9           for y in range(height):
10              gray_level = mat[x, y]
11              hist[gray_level] += 1
12
13      return hist
```

Code explained

This function takes an image as input, and returns the histogram representing the distribution of gray levels in it.

In line 6, we initialize a list of size 256. The i-th position in this list will hold the number of pixels in the image with the gray level equal to i. For example, the first position will hold how many black (0) pixels there are. In other words, we use the gray-level value (between 0 and 255) as an index.

This is how Otsu's method works: for every threshold t denote

$$
\begin{aligned}
back & = \text{number of background pixels } (<= t) \\
fore & = \text{number of foreground pixels } (> t) \\
mean\_back & = \text{mean value of the background pixels} \\
mean\_fore & = \text{mean value of the foreground pixels} \\
var\_between & = back * fore * (mean\_back - mean\_fore)^2
\end{aligned}
$$

The desired threshold is the one that maximizes $var\_between$ (which corresponds to the difference between the foreground and the background), among all possible thresholds t. Note that the value of $var\_between$ increases when the term $mean\_back - mean\_fore$ does, so the further away these means are, the better the threshold is. Also, the sizes of the foreground and background are taken into consideration: the value of the product $back * fore$ increases when these two variables are similar in size.

```
1   def otsu_thrd(im):
2       ''' return the optimal threshold for a 256 gray level image im '''
3       width, height = im.size
4       hist = histogram(im)
5       var_max = 0
6
7       for t in range(1,255):
8           back = sum([hist[i] for i in range(t+1)])
9           fore = sum([hist[i] for i in range(t+1,len(hist))])
10          if back == 0 or fore == 0: # an empty segment
11              continue # move to the next t
12          mean_back = sum(hist[i]*i for i in range(t+1)) / back
13          mean_fore = sum(hist[i]*i for i in range(t+1,len(hist))) / fore
14
15          # Calculate "Between Class Variance"
16          var_between = back * fore * (mean_back - mean_fore)**2
17
18          # Check if new maximum found
19          if (var_between > var_max):
20              var_max = var_between
21              threshold = t
22
23      return threshold
```

> ### Code explained
>
> This function first computes the image's histogram as we explained earlier. Then, it iterates over all possible thresholds `t` between 1 and 254 (note that we didn't bother to check `t=0` and `t=255`, as these would yield a `var_between` that is zero – make sure you understand why). For each possible threshold `t`, we compute `back` and `fore` by summing up the relevant parts of `hist` (until index `t`, and from `t+1` onwards).
>
> We then compute `mean_back` and `mean_fore` and `var_between` according to the formulas explained above. Finally, lines 19–21 store the best threshold so far.
>
> For example, consider a simple case in which there are only 10 pixels in an image. The variables *back* and *fore* may sum up to 10 in the following ways: $0+10$, $1+9$, $2+8$, $3+7$, $4+6$, $5+5$, $6+4$, $7+3$, $8+2$, $9+1$, $10+0$. The value of the product *back* * *fore* in each case is $0, 9, 16, 21, 24, 25, 24, 21, 16, 9, 0$. So it is maximal when both are 5. It is easy to generalize this and show that the multiplication of two values, whose sum is fixed, increases as these values are closer.

Now it's time for you to play with the code. Computing Otsu's threshold for the image in Figure 9.4, we get the following result. Try it! You can use this code:

```
>>> im = Image.open ("...")
>>> im = im.convert ('L')
>>> otsu_thrd(im)
154
```

> ### Exercise 2
>
> Find the best threshold (by Otsu's method) to segment the image in Figure 9.1 (available at our website).

9.2 Morphological Operators

In this section, we describe several common operators on images, termed morphological operators. The name comes from the Greek word "morph," which means shape. This is because these operators have one thing in common: they change certain pixels in the image as a result of the values of the pixels close by – that is, surrounding pixels. You can visualize this if you draw a continuous shape around a certain pixel, and look at the values of the pixels inside the shape in order to decide how to change the value of the "central" pixel. This shape defines the **neighborhood** of a pixel. For example, surrounding pixels can be those eight pixels that "touch" a certain pixel, which forms a neighborhood of a $3{\times}3$ square (see Figure 9.7), as in most of the examples in this chapter.

(a)

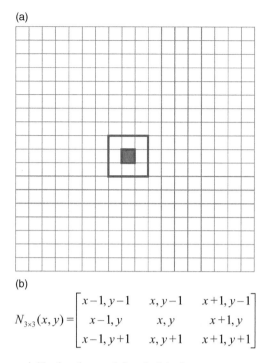

(b)

$$N_{3\times3}(x,y) = \begin{bmatrix} x-1,y-1 & x,y-1 & x+1,y-1 \\ x-1,y & x,y & x+1,y \\ x-1,y+1 & x,y+1 & x+1,y+1 \end{bmatrix}$$

Figure 9.7. A 3×3 square neighborhood around the pixel (x,y)
(a) Visual representation (b) Matrix representation

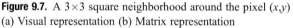

Figure 9.8. $(2k_x + 1)$-by-$(2k_y + 1)$ rectangle of pixels centered at (x,y)

More generally, we may be interested in a neighborhood that is different from a 3×3 square around a pixel. The neighborhood may be a square of larger sides, a rectangle, or any other shape containing the pixel. In this chapter, we will limit ourselves to symmetric rectangular neighborhoods. Symmetric means that the central pixel is at the center of the rectangle (see Figure 9.8). For any pixel (x,y) such a neighborhood can be represented by two parameters – k_x and k_y: k_x is the number of pixels to the left and to the right of (x,y) and k_y above and below it. For example, when $k_x = k_y = 1$, the neighborhood is a square of size

3×3; when $k_x = 1$ and $k_y = 2$, the neighborhood is a rectangle of height $2 \cdot 2 + 1 = 5$ and width $1 \cdot 1 + 1 = 3$ (the +1 in these calculations refers to the central row/column that contains the central pixel).

9.2.1 Erosion and Dilation

Erosion and dilation are two examples for important operations on images. Let us consider black–white images with white foreground and black background (such as Figure 9.9(a)). In a simplified manner, erosion turns any white pixel with at least one black neighbor into a black pixel (see Figure 9.9(b)). Dilation turns any black pixel with at least one white neighbor into a white pixel (see Figure 9.9(c)). In both cases, the affected pixels are normally those near the edges between foreground and background.

When the image contains only two levels of gray (in our case 0 for black and 255 for white), erosion can be performed by changing each pixel to the minimum of its neighbors. This way, if a white (255) pixel has at least one black (0) pixel in one of its neighboring cells, it will turn 0 (the minimum between 255 and 0). White pixels deep in a white area will remain unchanged. In the same context, dilation changes pixels to the maximum of their neighborhood.

> **Exercise 3**
>
> Are erosion and dilation opposite operations, in the sense that applying one operation on an image and then the other yields the original image?

Erosion and dilation use the same pattern of operation: applying some operator (minimum or maximum in this case) to each pixel's neighborhood, and assigning the result as the new value of that pixel. Since the only difference is in the actual operator taken, this calls for building a more general infrastructure: an abstract "**local operator**," whose instances will be the operations erosion and dilation (as well as additional operators, which we will see later on). Computer scientists call such a mechanism "high-order functions."

(a)   (b)   (c)

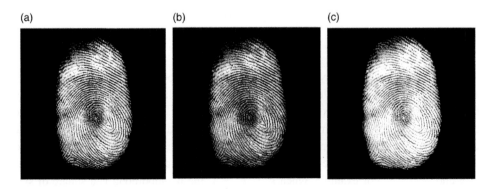

Figure 9.9. (a) Original image (b) Image after erosion (c) Image after dilation
Credit: MirageC / Moment / Getty Images.

Detour: high-order functions

High-order functions are functions that operate on functions. For example, a function may get another function as one of its parameters, or return another function when it terminates. In other words, a high-order function treats another (at least one) function as data. Until now, we were used to treating only types such as `int`, `float`, `str`, and `list` as data.

A good example for a high-order function that receives another function as a parameter is Python's `sorted`. When we call `sorted`, we give it a collection to sort, for example:

```
>>> sorted([1,-3,4,2,-5])
[-5, -3, 1, 2, 4]
```

But this function can receive another parameter, which specifies how to sort the list:

```
>>> sorted([1,-3,4,2,-5], key = abs)
[1, 2, -3, 4, -5]
```

The second parameter is a "named parameter", i.e., we call it using a name (`key` in this example), but this is not so important. What is important to see is that this parameter is a function! This function tells Python just how to compare the elements of the list to be sorted. In this example, the function is the absolute value function `abs` (another build-in function in Python). So during the sorting process, whenever Python compares two elements to decide in which order they should appear in the sorted list, the `abs` function is applied to these elements first. This is why -5 appears last in the output: $abs(-5) = 5$ and this value is the largest among all other absolute values of elements in the original list.

Here is another example. Recall Python normally compares strings by the lexicographical order:

```
>>> sorted(["hi", "hello", "hey"])
['hello', 'hey', 'hi']

>>> sorted(["hi", "hello", "hey"], key = len)
['hi', 'hey', 'hello']
```

In the second, call `sorted` compared strings by their length, not the lexicographical order.

We did not show here an example for a high-order function that returns a function, because this is less important for the rest of this chapter.

Back to erosion and dilation. The function `local_operator` uses an image `im`, an operation `op` (such as minimum or maximum), and the desired neighborhood parameters `kx` and `ky`. Since one of the parameters here is a function (`op`), `local_operator` is a high-order function. This function applies the operation `op` to each pixel's neighborhood.

```
1 def morph_operator(im, op, kx=1, ky=1):
2     ''' apply operator op on every pixel of im
3           with a neighborhood of kx and ky '''
4     w, h = im.size
5     out_im = im.copy()
6
7     mat = im.load()
8     out_mat = out_im.load()
9
10    for x in range(w):
11        for y in range(h):
12            left = max(0,x-kx)
13            up = max(0,y-ky)
14            right = min(w-1, x+kx)
15            down = min(h-1, y+ky)
16
17            xy_neighborhood = im.crop((left, up, right+1, down+1))
18            xy_neighbors = xy_neighborhood.getdata()
19            out_mat[x,y] = op(xy_neighbors)
20
21    return out_im
```

Code explained

The function traverses an image `im` and applies a given morphological operator `op` to each pixel in it. The neighborhood shape is determined by the horizontal (`kx`) and vertical (`ky`) lengths of the rectangle. The default is `kx` = `ky` = 1, which is a simple 3×3 square neighborhood.

Lines 12–15 compute the boundaries of the neighborhood. Normally, we would just need to go `kx` pixels to the left and to the right, and `ky` up and down. But we have to make sure we do not go beyond the image boundaries (left: 0, right: $w-1$, up: 0, down: $h-1$). Once we have the correct corners of the neighborhood, we crop it from the image in line 17, flatten this rectangular submatrix into a simple list in line 18, and apply the operator to this list of neighbors.

Now we can easily implement erosion and dilation. For the former, `op` will be Python's `min` function, and for the latter it will be Python's `max`:

```
1 def erosion(im, kx=1, ky=1):
2     return morph_operator(im, min, kx, ky)
3
4 def dilation(im, kx=1, ky=1):
5     return morph_operator(im, max, kx, ky)
```

Erosion and dilation are often used as part of, or as preliminary stages of, more advanced image-processing techniques. For example, the **opening** operator is defined as erosion followed by dilation. Opening smooth objects' contours, breaks the thin connections

(a)   (b)

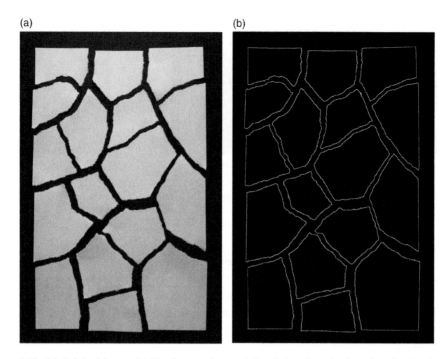

Figure 9.10. (a) Original image (b) The image after applying "edge detection" as described in the text. Credit: MirageC / Moment / Getty Images.

between them ("bridges") and eliminates small "islands." For example, opening can be used to break the thin connection between partly overlapping cells. The erosion step separates cells that were connected by thin bridges, and the following dilation step restores the original size of the eroded cells.

Similarly, **closing** is a dilation followed by erosion. Closing also smooths object contours, but merges objects with narrow gaps, and fills long thin gulfs and holes.

Erosion (or dilation) can also be used to extract the **edges** of objects in simple images. First, let us clarify what edges are. For most "normal" images, we expect a moderate change between a pixel and its neighbors. In other words, if we move from one pixel to its neighbor in any direction, we expect the brightness (or color, for color images) of the neighbors not to change dramatically. This is true for areas in the image that are **smooth**. However, in some areas of the image pixels do change extremely over small regions. Such areas are termed boundaries, or edges (see Figure 9.10). In biological images, edges may be the boundaries of a cell, the nucleus, various organelles inside a cell, etc. The term **piecewise smooth** is used to denote "generally smooth" images that contain edges. Edges often "summarize" an image in a compact (sparser) manner. Namely, they represent the important details in an image, using fewer bits of information.

Identifying the edges in an image is one of the most common tasks performed by those interpreting medical images. This is especially laborious when the structures in the image are of low contrast and minute details. Edge detection is a fundamental and important process in computer vision, and many clever algorithms exist for this.

What we show next is not really a method for edge detection, but rather a simple way to use erosion in order to draw the contours of well-separated objects. We first segment

(a)   (b)   (c)

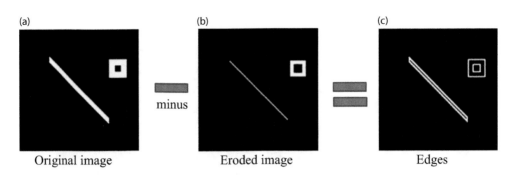

Original image   Eroded image   Edges

Figure 9.11. (a) The original image (b) The eroded image (c) The edges in the image

the image using Otsu's method. We then compute the difference between two matrices: that of the segmented image and that of the same image after erosion. The difference between the matrices is taken pixel-wise: we subtract from the value at each position (x,y) in the original image, the corresponding value at the same position in the eroded one. In the resulting image, positions that were 255 in the original image but 0 in the eroded image are given the value $255-0 = 255$. All the other positions are 0 (whether because they were both 255, or both 0 in the original and eroded image). The resulting image therefore contains the edges! (see Figure 9.11 for an illustration of this).

All that is left for us is to implement a simple `minus` function:

```
1   def minus(im1, im2):
2       ''' creates the difference matrix im1-im2 pixel-wise '''
3       assert im1.size == im2.size
4
5       w, h = im1.size
6       mat1 = im1.load()
7       mat2 = im2.load()
8       out = im1.copy()
9       out_mat = out.load()
10
11      for x in range(w):
12          for y in range(h):
13              out_mat[x,y] = mat1[x,y] - mat2[x,y]
14
15      return out
```

And finally the edge extraction function called `edges`:

```
1 def edges(im):
2     t = otsu_thrd(im)
3     seg = segment(im, t)
4     return minus(seg, erosion(seg))
```

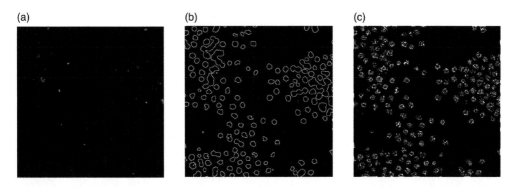

Figure 9.12. (a) Original image, of Drosophila Kc167 cells (b) Edge extraction using erosion (c) `PIL`'s edge detection

Try it!

```
>>> im = Image.open("...").convert('L')
>>> edges(im).show()
```

This simple method for edge extraction will work on very simple gray-level images, in which foreground and background differ extensively (that is, images that can be segmented well by binary thresholding). For most "real" images, this approach is too naïve. Python's `PIL` contains some standard edge detection methods, which are appropriate for various types of images. To use them, one applies a so-called "filter," which is a type of a morphological operator. For example examine the code below and then see Figure 9.12 for a comparison between the two methods.

```
>>> from PIL import ImageFilter
>>> pil_edges = im.filter(ImageFilter.FIND_EDGES)
>>> th = otsu_thrd(pil_edges)
>>> seg = segment(pil_edges, th)
>>> seg.show()
```

9.2.2 Noise Reduction (Denoising)

Digital cameras (as well as traditional film cameras) are susceptible to noise formation. Noise may result from a variety of sources, such as flecks of dust inside the camera, faulty camera or recording elements, or the deviation of electrons from their original path (a phenomenon called *electron hiss*).

At any pixel of the image, the noise component is defined as any deviation from the "true" value of that pixel. Denote the "true" value of the pixel as $T(x,y)$, and the noise component as $N(x,y)$, then:

$$M(x,y) = T(x,y) + N(x,y)$$

where $M(x,y)$ is the "observed" value of the pixel, that is, the value in the matrix that represents the image. There are many ways in which the noise component N can "behave". This is called the **noise model**. Typically, the noise component N is a random variable.

(a)   (b)   (c)

Figure 9.13. (a) Martin Luther King (b) Image after adding Gaussian noise (standard deviation $\sigma = 30$) (c) Image after adding balanced random salt and pepper noise (hit probability $= 0.02$)
Credit: Santi Visalli / Archive Photos / Getty Images

For example, one such model is the **Gaussian noise**, in which each pixel in the image will be changed from its original value by a small amount, which is distributed normally (in particular, small deviations from the original value are more likely than large ones) (see Figure 9.13(b)).

Another common noise model is the **salt and pepper noise**, in which sparse bursts of light and dark disturbances appear in the image (see Figure 9.13(c)). In brighter areas in the image, the dark (pepper) dots are observable, and in the darker areas the bright (salt) ones are seen. As a simplified example, think of a symmetric and random salt and pepper noise: every pixel has equal probability, denoted p, to get "hit", and change its value to complete white or complete black (each with probability $p/2$).

Given the observed image M, the goal of noise reduction, or denoising algorithms, is to produce a new image, which should be as close as possible to the original image T. Denoising cannot be achieved, if there are no constraints and assumptions on the image and on the noise. There are noise models that allow us to reconstruct either exactly the original value, or values very close to it. Other noise models may completely "erase" the original data, leaving no hope of recovering them even partially.

Let us clarify this. Suppose your friend chooses a number t between 0 and 255, then changes it arbitrarily by adding some other secret number n to it, also in the same range (0 to 255), and taking modulo of 256. Your friend only tells you that the new number $m = (t + n)\%256$ is 42. Can you find what the original number t was with high probability? Of course you can't. But suppose now you have heard that your friend only changed the original number by $n = +1$, n $= 0$ or $n = -1$. Now you have three possible guesses with equal probability ($\frac{1}{3}$) for guessing correctly: 41, 42, or 43.

There are many approaches to noise reduction, under various assumptions on the noise and on the image being cleaned. For example, there are **local** approaches that decide how to "clean" a pixel, based on its neighbors. Other, non-local methods, consider farther parts of the image. A famous type of a local denoising algorithm is **local medians**. This algorithm changes each pixel to the median of its neighbors. (See Figure 9.14. A median of a

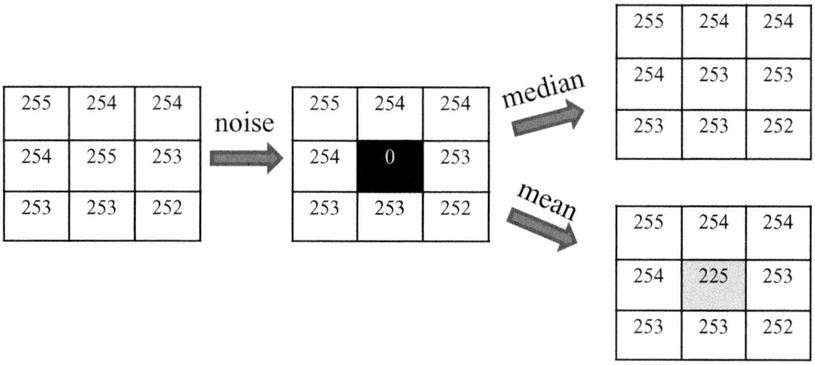

Figure 9.14. The difference between median and mean. The central pixel was changed from 255 to 0 (noise). The neighborhood is a 3 by 3 square. The median restores a value very close to the original one (253), while the mean (average) is 225, since it is more affected by the outlier pixel.

(a)   (b)   (c)

Figure 9.15. (a) Image with salt and pepper noise with hit probability 0.02 (b) After local medians (c) After local means

set is a value located at the middle of the sorted set. When the values in the set are unique, this is the value that is smaller than, or larger than, half the pixels).

Local medians perform particularly well in removing salt and pepper noise. This is because the median is not sensitive to extreme outliers, such as pixels with values of 255 or 0. Compare this with **local means**, where each pixel is replaced by the mean value (average) of its neighborhood (see Figure 9.14). A sudden "burst" of salt or pepper could affect this average dramatically. For reasons we will not explain here, local means is better for Gaussian noise, especially in smooth areas of the image.

Figure 9.15 shows the results of the two denoising approaches mentioned regarding an image with salt and pepper noise with hit a probability of 0.02 (in both cases, a 3×3 neighborhood was used for cleaning). Indeed the local medians approach cleans salt and pepper noise very well, with almost no deterioration in the quality of the image. With local means, salt and pepper noise is not removed and the image becomes blurred.

Like erosion and dilation, our two denoising operators are special cases of a local operator. As such, they can be implemented using the high-order function `local_operator`:

```
1  def mean(lst):
2      return round(sum(lst) / len(lst))
3
4  def local_means(im, kx=1, ky=1):
5      return local_operator(im, mean, kx, ky)
6
7  def median(lst):
8      return sorted(lst)[len(lst)//2]
9
10 def local_medians(im, kx=1, ky=1):
11     return local_operator(im, median, kx, ky)
```

9.3 A Biological Application – Cell Quantification

We complete this chapter with an example for a common biological use of image processing – counting cells. The goal is to replace a manual count, which is tedious and error prone. For that, we will use **labeling**, which we will now explain.

Labeling in our context is the identification of connected regions in an image. A connected region is a non-black area of pixels that are adjacent either horizontally or vertically. In our context, these areas are the cells in the image. Figure 9.16(a) shows a binary image with black and white pixels. Figure 9.16(b) shows its labeling – the labels are the integers 1,2,3,4, and 5. Each connected region of non-black pixels is assigned a number (0 is assigned for black pixels).

To perform labeling, we will not write our own code but rather use a very useful Python package called `scipy`. In order to work with `scipy`, you will have to install it first. Instructions and details appear at the website accompanying this book. One of the functions that `scipy` provides is `label`. Any non-zero values in the input image of `label` are considered

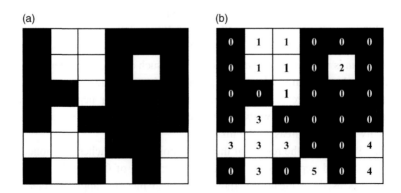

Figure 9.16. (a) A simple black and white image (b) Labeling of the left image: "objects" are labeled 1,2,3,4, and 5, while background is represented by 0

"features" or "objects" (for example, cells), and 0 values are considered the background. The function starts from every non-zero pixel, moves "out" to the neighbors of that pixel, and spreads until object boundaries are reached in all directions (so the whole object is "covered"). The objects are labeled using the integers 1,2,3,..., while pixels that were not identified as part of objects remain 0. So in the end each unique object or feature in the input image has a unique (positive integer) label.

Here is the code that labels the image from Figure 9.16:

```
>>> from scipy.ndimage.measurements import label

>>> mat =[[  0,255,255,   0,   0,   0],
          [  0,255,255,   0,255,   0],
          [  0,  0,255,   0,   0,   0],
          [  0,255,  0,   0,   0,   0],
          [255,255,255,   0,   0,255],
          [  0,255,  0,255,   0,255] ]

>>> labeled_array, num_features = label(mat)
>>> print(num_features)
5
>>> print(labeled_array)
array([[0, 1, 1, 0, 0, 0],
       [0, 1, 1, 0, 2, 0],
       [0, 0, 1, 0, 0, 0],
       [0, 3, 0, 0, 0, 0],
       [3, 3, 3, 0, 0, 4],
       [0, 3, 0, 5, 0, 4]])
```

The `label` function returns the number of objects found (5 in this example), and the labeling itself as a matrix. Note that neighboring pixels, through which the algorithm can "spread out" from the initial pixel, do not include those that touch it only diagonally. One of the parameters of the function `label` allows changing this definition of the neighboring pixels, and this is called a "structuring element" (not shown here).

Now let us use this function to count how many cells are in Figure 9.17 of human colon cancer cells.

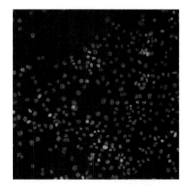

Figure 9.17. Human color cancer cells. How many cells are in that image?
Source: https://data.broadinstitute.org/bbbc/BBBC001/image AS_09125_050118150001_A03f00d0

First, we apply Otsu segmentation to that image, for otherwise almost all the pixels will be considered objects, as they are normally above 0. (How many labels are we expected to get in this case?)

```
>>> im = Image.open("…")
>>> th = otsu_thrd(im)
>>> th
38
>>> im2 = segment(im, th)
>>> from scipy.ndimage.measurements import label
>>> mat, n = label(im2)
>>> n
367
```

Applying `scipy`'s labeling algorithm yields a count of 367 cells.

Of course, this quantification may deviate from the truth because segmentation may have eliminated relatively dark cells, as well as because some cells are too close for the labeling to count separately ("clusters" of cells). More sophisticated methods for cell quantification are beyond our scope.

Reflection

In this chapter, we explored the basics of digital image representation and processing. Digital images are represented as matrices, whose entries correspond to image pixels. These entries hold numeric values that indicate the brightness of the pixel: a single value for gray-level images, and three values for color images (e.g., one value for each of the RGB components).

We introduced several basic image-processing techniques. Segmentation is a fundamental task in image processing. We showed how segmentation can be achieved through thresholding, but this is only a very basic segmentation approach, and other approaches exist. Dilation and erosion are two fundamental local operators that have many uses. Here, too, there exist other, much more sophisticated, methods for detecting edges, and we have merely scratched the surface. In fact, edge detection is sometimes used as the first step in segmentation (and vice versa, as we saw). We note that the visual neural system in our brain performs edge detection at the early stage of analyzing images.

The implementation of functions in this chapter is an example for what programmers call "modular software design." This means that code is not duplicated, and divided into modules or parts in a hierarchical structure (see Figure 9.18).

Modular design has several advantages. As mentioned, it prevents code duplication, it allows easier testing, and it supports parallelization of development (i.e., several teams working on parallel modules at the same time). In this chapter, we defined an abstract morphological operator function, which could be invoked with various operators, such as minimum (erosion), maximum (dilation), mean, and median. There is usually a tradeoff between implementing the just-needed functions, versus preparing the whole infrastructure for future extensions of the code. This is part of what software developers are occupied with.

Image processing goes way beyond the fundamental operations introduced in this chapter. One example is the problem of *tracking*. Tracking aims at locating moving objects in

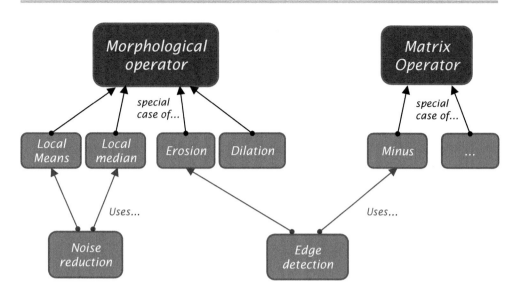

Figure 9.18. Modular software design illustration

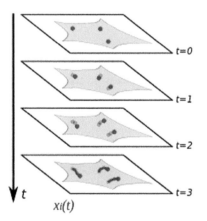

Figure 9.19. Image (and following text) from Wikipedia. Principle of single particle tracking: The rectangles represent frames from an image acquisition at times $t = 0, 1, 2, \ldots$. The tracked particles are represented as red circles and in the last frame, the reconstructed trajectories are shown as blue lines.

consecutive images taken over time (such as video frames). This can be especially difficult when the objects are moving fast relative to the frame rate, and when the tracked objects change orientation or shape over time, disappear from the frame, or divide into several objects (e.g., cell division) (see Figure 9.19).

As we have witnessed in this chapter, some tasks may still require human intervention. One of the reasons for that is the high diversity in imaging technology. For example, fluorescence microscopy produces images very different in nature from MRI images. Another reason is the high diversity in image content: an edge detection algorithm may be tailored for specific shapes such as elongated cells (rather than round or elliptic ones). Identifying cell centers may be performed by locating local maxima of brightness, but this assumes central regions of cells have higher intensities that gradually fall towards cell boundaries, etc.

In this chapter, we used Python's `PIL` imaging package, as well as the `scipy` package. These packages contain many more functionalities than we exposed here, and there are other packages as well. Fortunately, these all have a rich and informative documentation, as well as supporting forums, and thus are relatively easy to learn and use. Obviously, some well-known existing tools, such as Photoshop and others, enable many image-processing tasks to be performed without writing a single line of code. However, this chapter's goal is to expose the readers to the interiors of the world of digital images. In addition, the commercial tools and packages may lack the specific functionality that one may need, and so the hands-on experience learned in this chapter should come in "handy" in the future.

Image processing is mainly investigated in academia in the departments of computer science and of electrical engineering. Subfields of this area are *computer vision* (which aims at computerizing human vision aspects, such as facial recognition, and involves some form of intelligence and cognitive processing), and *computer graphics* (referring to image data created by a computer).

Challenge Yourself

Problem 1 Cell quantification

On the website, you will find the Figure 9.20. It contains an image of Drosophila Kc167 cells. Give your best estimation for the number of cells in this image. Use Python's labeling to answer this. You may use any image-processing technique to improve the results, such as dilation, erosion, noise reduction, and segmentation.

Figure 9.20. Drosophila Kc167 cells

Part V: Limitations of Computing

– "Easier said than done."

Computers and computer programs are harnessed these days to solve an extremely diverse range of problems. These problems include weather forecast, scheduling problems, designing, and controlling trajectories for satellites, recognizing individual human faces, playing chess, trading in the stock market, diagnosing cancer cells, assembling automobiles and airplanes, and driving autonomous cars (this is obviously a very partial list). In many cases, computers outperform human experts working on the same problem. In the future decades, we may expect computers to gradually replace medical doctors (completely or partially), express emotions, compose high quality music, etc. etc. Given these amazing fits, one wonders if computers are omnipotent: Can computers solve every problem, provided it is well defined?

We note that certain problems are not computational in nature and one cannot expect computers to be any better than humans in approaching them. For example, questions such as characterizing beauty, the existence of god, and whether human nature is basically good, as claimed by Jean Jacques Rousseau. But suppose we restrict ourselves to computational problems that can be formally described using a precise mathematical definition, and have a well-defined yes/no answer. One may think that a talented programmer could, in principle, devise an algorithm to solve any such problem, given enough time (and financial resources...). But – is this indeed the case?

The answer to this fundamental question is negative. Whether this is good or bad news is debatable. But understanding the capabilities and limitations of computing is an important intellectual challenge, which also sheds light on what to expect, and what to avoid, in working with computers. Interestingly enough, the precise, mathematical claim that computers are not omnipotent, but are limited in a very fundamental way, was proven in 1936. This was well before digital computers existed. This result was proved by Alan Mathison Turing, the British computer scientist, mathematician, and logician.

In the last two chapters of the book, we introduce the limitations of computing. In chapter 10, we will discuss problems that cannot be solved at all, while in chapter 11 we will describe some hard problems that can be solved in principle, but for which any known solution is too expensive in terms of running time.

10 Mission Impossible

In this chapter, we will define Turing's **Halting Problem**. We will then describe Turing's proof that the Halting Problem cannot be solved by any computer, ever. This was the first problem proven to be, fundamentally, beyond the capabilities of any computer. This result subsequently served as a springboard to prove that a whole, infinite class of other problems are unsolvable as well.

10.1 The Halting Problem

Consider the following function, `stoppy`.

```
1 def stoppy(n):
2     i=1
3     while i<=n:
4         i=i+1
5         print(i)
6     return "hello, world!"
```

It receives an input, n, which is assumed to be an integer. It prints all integers in the finite range starting from $i = 1$ and ending at $i = n$. Upon leaving the loop, the program returns the well-known phrase "hello, world!". If $n < 1$, the code loop terminates immediately, and no integer is printed.

```
>>> stoppy(8)
1
2
3
4
5
6
7
8
'hello, world!'
```

One author of this book (whose identity will be kept confidential!) set out to rewrite the program, but he did so after consuming noticeable quantities of high quality beer. The resulting function, which we name `loopy`, is almost identical to `stoppy`, except the line $i = i + 1$, which was inadvertently dropped (rumor has it this is due to the influence of beer).

```
1 def loopy(n):
2     i=1
3     while i<=n:
4         print(i)
5     return "hello, world!"
```

```
>>> loopy(8)
0
0
0
0
# aborted manually
```

In "real life," such small changes usually have very small impact. But here the impact is dramatic. When the program is executed with $n \geq 1$, the loop is entered. The condition $i \leq n$ remains valid in each iteration, as neither i nor n is changed. As a result, the execution flow never leaves the loop. We are facing what is known as an **infinite loop**, or, equivalently, a **non-terminating program**. The only way to get out of this infinite loop is to abort it manually (for example, if you use IDLE, by pressing Restart Shell (CTRL+F6) in the IDLE menu).

The two functions, `stoppy` and `loopy`, are so simple that we can easily analyze their behavior, even without the help of existing formal verification tools (these tools are extensively used in the software and hardware design industries). However, most programs we write and then run are not as simple as these two toy programs. It is very important for the programmer to be certain that her program will eventually halt. If a programmer knows that on a certain input, the program she wrote either runs for two days and then halts, or never halts, she could abort the execution once two days had elapsed. But suppose nobody knows whether the program is going to halt or not, and the program has already been running for six days and has not halted. If the program is not going to halt, we may as well abort it, save energy, and free our expensive computing resources for more fruitful tasks. But if the program is going to halt with the sought solution after seven days of execution, and we have aborted it after six and a half days, we will have wasted expensive resources, and tossed an almost complete solution into the garbage bin.

Formally, the halting problem is the following: given a program, f, and an input to it, x, does f halt when executed with input x?

Exercise

Fast Fox is one of the smartest students in class. While he certainly respects Turing and his class lecturer, he believes the halting problem is easy to solve, and suggests the following solution:

Given f and x, run f on x. If it halts, return the answer "f halts on x".

What is wrong with Fast Fox solution? Where does it fail to solve the halting problem?

Programs need not be overly complicated or have many lines of code to pose a challenge related to halting. The following, known as the as the **Collatz process** (after the German mathematician Lothar Collatz), provides an illuminating example.

10.2 The Collatz Conjecture

The process begins with a single input, which can be any integer $n \geq 1$. Given the current number, the next one in the sequence is defined as follows: If the current number is even, we divide it by 2, and if the current number is odd, we multiply it by 3, and then add 1. This process terminates once we have reached the number 1. A simple Python function, implementing the Collatz process, is given below.

```
1   def collatz(n):
2       ''' runs the Collatz iteration and prints each element till
3           1 is reached. Will not halt if 1 is never reached '''
4       initial = n
5       print(initial)
6       while n!=1:
7           if n%2 == 0:
8               n = n//2
9           else:
10              n = 3*n+1
11          print(n)
12
13      # if we got here, n must be 1
14      print("halted on initial input," initial)
15      return
```

The function prints all intermediate values of the variable n, which are generated according to the Collatz process. If the variable n reaches the value 1, the function leaves the main loop, prints the initial input, returns None, and halts.

Let us examine the execution of this function on several inputs, focusing on the question of halting:

```
>>> collatz(64)
64
32
16
8
4
2
1
halted on input 64
```

The intermediate values, printed by this function, are 64,32,16,8,4,2, and then 1, leading to halting. Observing this sequence, we see that 64 is a power of 2, and thus all intermediate values are also powers of 2. The function keeps dividing by 2, and no odd number is encountered before reaching 1 and halting. It is easy to see that *every* power of 2 exhibits a similar behavior.

Exercise

(a) Try this out yourself, e.g., running `collatz(512)`, `collatz(1048576)`, `collatz(2**30)`, or even `collatz(2**50)`. Note that $2^{50} = 1,125,899,906,842,624$ is a fairly large number. Yet we claim that the number of iterations till 1 is reached will be small enough for your computer to finish execution fast. Why is this so? Can you tell beforehand how many iterations the function will execute on 2^{50}?

(b) Interestingly, the number of steps needed to reach 1 does not grow monotonic-ally with the input n. Try $n = 13$, $n = 27$ to see that.

(c) We want to check whether Collatz halts on a given range of inputs. Suppress all the intermediate printing from the function `collatz`, and write a function `check_collatz(k)` that calls it on all integers from 2 to `k`. Note that `check_collatz` will halt if and only if `collatz` will halt on all inputs from 2 to `k`. Call `check_collatz` with `k = 1,000,000`. Did it halt? Can you conclude from this experiment that the collatz process terminates for *every* integer n?

While understanding how the `collatz` behaves on n which is a power of 2, other values of n do not exhibit such simple behavior. For example, $n = 9$ exhibits the following dynamics:

```
>>> collatz(9)
9
28
14
7
22
11
34
17
52
26
13
40
20
10
5
16
8
4
2
1
halted on input 9
```

While for $n = 13$, we have

```
>>> collatz(13)
13
40
20
10
5
16
8
4
2
1
halted on input 13
```

These two input values demonstrate that the behavior of the Collatz process is hard to predict. Intermediate values can go up as well as down, making it hard to tell how many iterations will be executed on a specific input. But there is an even more fundamental issue here: it is hard to even determine *if* the Collatz process halts or does not halt on a specific input. We can always run the code on any given n, but while running we will not know if it is going to halt or not.

A famous conjecture in mathematics, termed the **Collatz conjecture** or the **3n + 1 conjecture**, states that for *every* integer $n \geq 1$ the Collatz process is finite. In other words, the conjecture states that the function `collatz` will eventually halt for every such n. This conjecture was tested computationally by many, as of 2019 reaching values of n as large as $87 \cdot 2^{60} = 100{,}304{,}170{,}900{,}795{,}686{,}912$. On all these values, the function halted. Note that this finding may increase our confidence that the conjecture is true, but it by no means proves it.

Exercise

(a) Suppose you suspect some integer k refutes the $3n + 1$ conjecture (that is, when the process starts with k, it will never reach the number 1). Explain why running the function `collatz` with this input, k, will not help you disprove the $3n + 1$ conjecture.

(b) Consider the following function (which has no input):

```
def check_all_collatz():
    n = 2
    while True:
        collaz(n)
        n = n+1
```

Explain why the `check_all_collatz` function cannot assist us in proving or disproving the $3n + 1$ conjecture.

(c) If the $3n + 1$ conjecture were false, there must exist an integer m for which the Collatz process does not terminate. Consider the sequence of intermediate values produced during the execution on the input m. This sequence is either unbounded, or bounded. In the latter case, how can you detect this scenario, using a small modification to the code of the existing `collatz` function?

10.3 The Halting Problem Is Undecidable

While the interest in the Collatz process is mainly mathematical, halting of general programs is a very serious issue, faced by millions of programmers, worldwide. You may have got the impression that computers are omnipotent: computer programs solve all our problems, and are expected to replace us in almost any mathematically defined task altogether in the near future. So, maybe all that is required is a talented programmer who will write a program to solve different variants of the halting problem. In the Collatz case, we would like to determine if the program halts on every positive integer input, n.

It would be great if this was the case, but unfortunately, it is not. In 1936, Alan Turing has shown that the halting problem is **undecidable**. This means that nobody can write a program, call it *Halt*, that takes as inputs another program, f, and an input to that other program, x, and determines if f halts when executed on x. By Turing's result, any such program, Halt, will either make errors, or enter infinite loops itself! The inability to write such a program does not follow from the incompetence of past, present, or future programmers, from the lack of expressive power of the programming languages we use, or from some basic shortcoming of our processors. Instead, this is an inherent, fundamental limitation of computing.

How did Turing prove such a bold statement, at a time when digital computers did not even exist? The mathematical approach is what is known as **proof by contradiction**. Assume that such a program, *Halt*, does exist, and show that by running *Halt* on a simple modification of itself, we get a program that both halts and does not halt on the same input, which is of course impossible.

To prove this claim, Turing used an abstract model for what a computer is, which we call today the **Turing machines**. This model is much simpler than modern computers, whose full hardware and operation details are probably even beyond the grasp of most experts. Incredibly enough, in principle, Turing machines can do anything a modern computer can do (albeit less efficiently). So in terms of computational capabilities, Turing machines can accomplish everything an expert programmer, equipped with the strongest computer on earth, can do (regardless of the programming language used). In other words, if a problem is not solvable by a Turing machine, then no computer program can solve it.

Turing showed that some problems, and specifically the halting problem, cannot be solved by Turing machines, and thus not by any computer. We will prove this important impossibility result using the terminology we are already familiar with, that of Python programs and functions. We note that these are equivalent to Turing machine, and thus to any computer program, as well.

Let us now turn to a formal proof that the halting problem is not solvable by any computer program. Suppose, as a contradiction, that there is a Python program, Halt, which solves the halting problem:

```
1  def Halt(f,x):
2      ...
```

This program has two inputs: the first is another Python function (or program) of one input parameter, f, while the second, x, is an input to f. We remind you that having a function serve as an input to another function is a standard practice in Python, and in many other programming languages as well, termed high order functions.

Saying that Halt(f,x) solves the halting problem means that on inputs f and x, Halt(f,x) returns True if f halts on x, and False otherwise (that is, when f enters an infinite loop when executed on x). In particular, Halt itself halts on *every* pair of inputs f,x. Note that if running f on x leads to an error (for whatever reason – incorrect format, overflow, wrong type, division by 0, etc., etc.) we consider this as a halting of f on x.

We now construct a new function, which we term `modify`. This function employs `Halt` in an essential fashion. The function `modify` takes as an input a function, `f`.

```
1 def modify(f):
2       if Halt(f,f):           # f halts when given f as input
3           while True:         # run forever!
4               x=0
5       else:                   # f does not halt on input f
6           return "yes, we can!"
```

The function `modify` starts by calling `Halt(f,f)`. Note that there is nothing unusual about this: `f` can get any kind of input, including `f` itself (more specifically, the Python code of `f`). If `f` halts when run on itself as input (either because this yields an error or because the execution terminates smoothly, then `Halt(f,f)` returns `True`. Otherwise, if `f` enters an infinite loop when run on itself, `Halt(f,f)` returns `False`. Note that since `Halt` itself is assumed to halt on every input, the line in the code `Halt(f,f)` will always terminate, so this condition will never lead to an infinite loop.

Now consider what will happen if we apply `modify` to *itself*, namely invoke `modify(modify)`.

```
>>> modify(modify)
???
```

Specifically, will it halt or not? Like any execution of a program on some input, the execution `modify(modify)` either halts or does not halt. To make the presentation clear, let us explicitly look at the code when we substitute `modify` for `f`:

```
1 def modify(modify):
2       if Halt(modify, modify): # f halts when given f as input
3           while True:           # run forever!
4               x=0
5       else:                     # f does not halt on input f
6           return "yes, we can!"
```

We examine both options, and show that each of them leads to a **contradiction**:

1. If `modify(modify)` does not halt, this means it entered the infinite loop in lines 3 and 4. But this can happen only if the condition in the `if` was satisfied, meaning `Halt(modify, modify)` returned `True`, which says `modify` does halt when run on itself as input.

2. If `modify(modify)` halts, this means it returned `"yes, we can!"`, which can happen only when `Halt(modify, modify)` returned `False`, which says `modify` does not halt when run on itself as input.

We showed that `modify(modify)` halts if and only if `modify(modify)` does not halt. This is a contradiction!

Specifically, this is a contradiction to the existence of a program `Halt(f,x)`, that solves the halting problem. This means that the halting program can never be solved by any computer program.

10.4 Reflections on the Proof

This proof resembles what you may know as the "**liar's paradox**": the case of a person declaring "every statement I make is a lie." If this is true, then the statement is not a lie. If it is false, then the statement is true. So both options lead to a contradiction. The idea of having a program that relates to a modified version of itself is a fundamental idea in mathematical logic. The other two best-known examples of this idea are probably the following:

1. Georg Cantor (1845–1918) proved that there are more real numbers than integers (even though there are infinitely many reals and integers, so there is more than one "infinite"). The proof is by a so-called *diagonalization* technique.
2. Kurt Gödel (1906–1978) proved that there are statements in the language of number theory so that neither the statement nor its negation are provable within the axioms of number theory. This is known as *Gödel's incompleteness theorem*. It was proved by a sophisticated encoding of statements as numbers, enabling the formation of a statement saying "I am not provable".

Both results and their proofs are fundamental results in mathematics.

11 Mission Infeasible

In the previous chapter, we studied well-defined computational problems that cannot be solved by any computer program. Consider now a setting where we are given a well-defined computational problem that is solvable by some computer program. Yet any such program takes a very long time to complete. Long enough that by the time the execution terminates, the solution is no longer relevant.

In this chapter, we briefly introduce the subject of **computational complexity**. In particular, we will develop some intuition for computational problems that have efficient computational solutions, versus those problems that are computationally infeasible (despite being solvable in principle).

These notions are of great theoretical importance in computer science, and the attempts to understand them are central to understanding the foundations, and limitations, of computing. But, in addition to their theoretical aspects, they have important practical aspects as well. When faced with a new problem, being able to classify it as having an efficient solution, versus one that is computationally infeasible, will help understand when a solution is within reach and when it is not. In the latter case, a number of alternative approaches may be viable. For example, an approximation algorithm, which yields a solution that is "close enough" to the exact solution.

11.1 Measuring Computational Complexity

Let's start with a short reminder on complexity, as described in Chapter 2. When analyzing the computational complexity of a problem, it makes no sense to restrict ourselves to a single, fixed problem instance. This would have a fixed answer, with few implications for other instances of the general problem. Instead, we measure the computational complexity as a function of the input size. For example, if we want to add two n bit integers, and use the simple algorithm taught in primary school, the complexity of this algorithm is *linear* in n. If we want to multiply two n bit integers and use the simple algorithm taught in primary school, the complexity of this algorithm is *quadratic* in n. Moving from integer arithmetic to graph algorithms, let us recall the breadth first search algorithm (BFS) on a non-weighted, connected graph. Here, we have two relevant input parameters: the number of nodes, n, and the number of edges, m. The algorithm, as described in chapter 6, visits each node and each edge exactly once. The complexity of this algorithm is linear in $n + m$.

Let us go back to algorithms with a single input parameter, n. Denote the number of steps taken by some algorithm on inputs of length n by $T(n)$. If there are constants a, b, c such that for all n, and for all inputs of length n, $T(n) \leq an^c + b$, we say that the (worst case) running time of the algorithm is **polynomial** in the input length. Furthermore, we often summarize the running time by just the exponent, c. For example, $c = 1$ corresponds to linear time algorithms. The algorithm for computing the GC contents of a genomic string is linear (in the string length). When $c = 2$, we say the algorithm has quadratic run time. When $c = 3$, we say the algorithm has a cubic run time, and so on.

Consider the following algorithm for checking if all elements in an n long list are distinct: compare the first element to the last n-1 elements; compare the second one to the last n-2 elements; and so on, till the last two elements are compared. If at any moment we encounter two equal elements, return False. Otherwise, return True at the end. Overall, the number of comparisons is

$$(n-1) + (n-2) + \cdots + 1 = \sum_{i=1}^{n-1} i = \frac{n(n-1)}{2} = O(n^2)$$

Note that this expression is indeed quadratic in n and that we used the O notation presented in Chapter 2.

We make two important remarks related to the above definition and examples. Typically, most algorithms need to read their entire input, and thus the number of operations they will require is at least linear in the input length. One notable exception is binary search, where we assume the input list is already sorted, and can search for the desired key *without* reading all input items. Second, the fact that an algorithm for some problems runs in a certain time complexity does *not* imply that the problem cannot be solved by a more efficient algorithm. Consider, again, the distinctness problem, for which we described a very intuitive, quadratic time solution. It is possible to devise a more efficient algorithm to this problem: sort the list, and check if after sorting there are no two identical consecutive elements. The complexity of this solution depends on the method we choose for sorting. There are $O(nlogn)$ sorting algorithms we can use for that. In addition, going over the sorted list takes $O(n)$ time. Thus, in total this alternative approach to solve the distinctness problem takes $O(nlogn) + O(n) = O(nlogn)$ time.

Unfortunately, not all computational problems that are solvable have polynomial time solutions. There are problems requiring, for example, $T(n) = 2^n$ or even $T(n) = n!$ steps on inputs of size n, where $n!$ is the **factorial** function, $n! = 1 \cdot 2 \cdot \ldots \cdot n$. Algorithms whose running times grow at such rates are called **exponential** time algorithms. In this chapter, we will see a couple of examples for computational problems with such solutions. For large inputs, polynomial time algorithms are preferable to exponential time ones, as the actual time to completion taken by a polynomial time algorithm will be way shorter than that taken by an exponential one.

11.2 Maps and Coloring

Consider a simple map of areas consisting of different, connected regions (these could be countries, states, municipalities, etc.). We refer to this as a **planar map** (map on the plain). To improve the readability of such maps, we usually color the different regions. A coloring is called *legal* if regions that have a common border are colored using different colors. Legal coloring enables us to easily tell apart different regions. Figure 11.1(a) shows a legal coloring of a specific map, using just two colors. We emphasize that a single point common to two regions does not constitute a border. To qualify, a border must consist of a continuous line, shared by the two regions. Figure 11.1(b) also shows a map that is colored using three colors. This map contains three neighboring regions, thus the map requires at least three colors. If a map has a legal coloring using k colors we call it **k-colorable**. The map in Figure 11.1(a) is

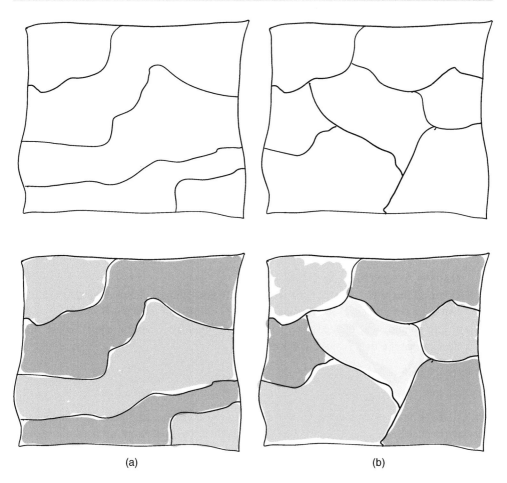

Figure 11.1. (a) A 2-colorable map and a possible 2-coloring (b) A map that cannot be colored using just two colors, and its coloring using three colors

2-colorable. It is also 3-colorable since we can also use three colors instead of two (simply color any arbitrary region red for example). The map in Figure 11.1(b) is 3-colorable (and also 4-colorable, 5-colorable, etc.), but as we mentioned it is not 2-colorable.

Consider the following problem: Given a planar map, is there a legal coloring of it, using just *two* colors? In other words, is it 2-colorable? Such a problem is called a **decision problem**, since the answer is either "yes" or "no". Suppose the map has n regions, and furthermore, that our favorite colors are red and green. Each region can be colored either red or green, so there are two possibilities per region. For n regions, there are 2^n ways to color them. We could now go over each such coloring, and check if it is a legal one. If it is, the answer to our problem is "yes" – the map is 2-colorable. If none of these colorings is legal, then the answer is "no," and the map is not 2-colorable. Note that a positive answer requires just one legal coloring, while to reach a negative answer we must in principle go over all possible colorings using two colors and see that none is legal.

The algorithm described above takes exponential time, in particular it requires in the worst case (when the map is not 2-colorable) $o(2^n)$ steps. For $n = 10$, this is just fine, and even for $n = 20$, the algorithm will complete in less than one second on a modern computer. But for $n = 100$, the algorithm requires that we go over $2^{100} = 1267650600228229401496703205376$ colorings, which is way too many! Even with extensive computational power, this will not terminate by the end of the 21st century, or even before our sun will run out of energy.

The fact that the naïve algorithm, described above, is too slow, does not imply we cannot do better, using a more efficient algorithm. In this case, the following simple algorithm does solve the 2-coloring problem efficiently:

Checking if a given map is 2-colorable

Start with any region (e.g., the one closest to the lower left corner of the map) and color it green.
Next, iterate the following:

- Pick one of the most recently colored regions. Suppose it was colored by color A (for example A = green).
- Color all the direct neighbors of it (namely those having a common border with it) that were not already colored using the other color B (for example B = red).
- If any conflict was created (two neighbors colored by the same color), abort and declare that the map is not 2-colorable.

Declare the map is 2-colorable

If the process ends and was not aborted, the map is 2-colorable, and in fact the coloring produced is a legal 2-coloring of the map. Note that this algorithm does not require $o(2^n)$ time, and in fact its complexity is polynomial in the number of regions in the map. Great!

What about deciding 3-colorability?

Given a map with n regions, we can try all 3^n possible colorings, and check if any of them are legal. Clearly, 3^n is way larger than 2^n, so this algorithm will not be very practical for large values of n. When faced with a similar situation regarding 2-colorability, we were able to devise an efficient algorithm to solve the decision problem, and even find a legal coloring, if one exists. Why not go through a similar process for 3-colorability?

Well, this could be nice, but nobody has found a polynomial time algorithm that determines if a given planar map is 3-colorable. The main difference is that for 2-coloring, we got only a single option of coloring a region bordering one that was colored already (we must choose the other color). For three coloring, we got two options (the other two colors). This causes a so-called "combinatorial explosion," forcing the algorithm above to explore two (rather than one) options at every branching point, and exponentially many options overall. So while determining 2-colorability can be done efficiently, it is unknown whether 3-colorability also has an efficient solution. As we explain later, most computer scientists believe this problem does *not* have an efficient solution. This belief has no proof, however, and it may never have one.

Let us now turn to the next step in this sequence, that of 4-colorability. It is easy to produce maps that are 4-colorable but not 3-colorable. One example is shown in Figure 11.2.

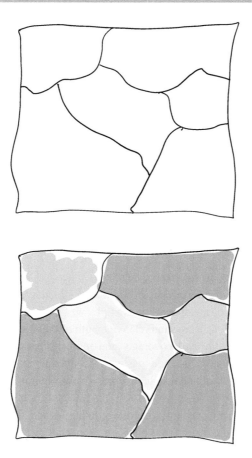

Figure 11.2. A map that requires four colors to map. It is 4-colorable but not 3-colorable

---

🖊 **Exercise 1**

Can you provide a simple explanation why the map in Figure 11.2 is not 3-colorable?

On to the next one: is there a planar map which is *not* 4-colorable? This problem, known as the **4-coloring problem**, has an interesting history.

In the year of 1852, Francis Guthrie was a 21-year-old mathematics and botany student at University College London. While trying to color the map of the counties of England, he realized that three colors are not enough, but four colors are. From this observation, regarding a single map, Guthrie made a bold generalization, known as the **4-color conjecture**. It states that *every* planar map has a legal coloring using (at most) four colors. The extreme simplicity of the statement of the 4-color conjecture encouraged many mathematicians to try and prove it. However, the proof turned out to be far more elusive than expected.

A number of incorrect proofs were proposed, and it was only in 1976 that the conjecture was finally resolved. It was proved by Kenneth Appel and Wolfgang Haken, who showed that if each of over 1936 specific maps are 4-colorable, then every planar map is. To verify that each of these 1936 maps are indeed 4-colorable, Appel and Haken applied a computer program that checked each of them. The use of computers in mathematical proofs was not universally accepted, mostly because a bug in the program could go unnoticed. Later, their approach was simplified, and the program correctness was verified by general-purpose theorem-proving software. This has increased the confidence in the correctness of the **4-color theorem**, which now is very high.

The 4-color theorem resolves the 4-colorability decision problem with a "yes" answer for any planar map. In other words, given the theorem, we know that every planar map is 4-colorable. Furthermore, for those interested in finding an actual 4-coloring, not just knowing that one exists, a polynomial time algorithm was devised by Robertson, Sanders, Seymour, and Thomas in 1996. It is fairly complicated, hence we will not give its details here.

11.3 From Maps to Graphs

In this section, we refer to an interesting relation between maps and undirected graphs. A regular map conveys lots of information, in addition to which regions are neighbors (have a common border) and which ones are not. For example, we get information on the shape of the different regions, on their relative area and relative length of their boundaries, and possibly on the topography, the roads, the land coverage, and more. Suppose that all this additional information is of no interest to us, and we want to abstract it away, keeping only the neighborhood information. It turns out we can represent exactly this information, and nothing else, using graphs.

Given a planar map, we build a corresponding graph. Each node of this graph corresponds to a region in the map. Two nodes are connected by an (undirected) edge if and only if the two regions have a common border (neighbors in the map). See an example in Figure 11.3.

Graphs can also be colored, just like maps. A coloring of a graph is an assignment of colors to its nodes, one color per node. A coloring of a graph is called legal if any two nodes that have an edge between them are colored by different colors. Obviously, if a map can be colored using k colors, so can the corresponding graph. However, these problems are not

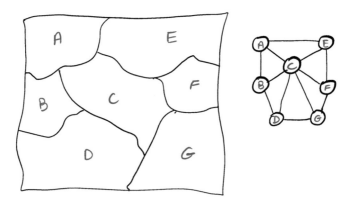

Figure 11.3. A map and its corresponding graph. Each region is a node, and two nodes are connected with an edge if and only if the corresponding regions have a shared border (that is not a single point)

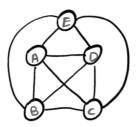

Figure 11.4. An example for a graph that does not represent any planar map

equivalent: for every map we can produce its corresponding graph, but not vice versa. Take for example the graph in Figure 11.4 – it is not possible to draw a corresponding planar map – try it!

Analogously to the question of map coloring, when given a graph G on n nodes, we can ask: Is it 2-colorable? 3-colorable? 4-colorable? ... k-colorable? ... n-colorable? Well, the answer to the last question is always positive: By simply assigning a different color to every node, the resulting coloring is legal. At the other end of the line, the question of two colorability can be answered by a simple algorithm, pretty much the algorithm for maps, described earlier.

But what happens for k-colorability with k greater than 2? The uncomfortable truth is that we do not know any efficient algorithm to resolve the graph coloring question, for any $k > 2$. Despite over 40 years of extensive research in Computer Science, Operation Research, and Combinatorial Mathematics, we do not know the answer. On one hand, nobody has found an efficient algorithm for this task. On the other hand, despite substantial effort, no mathematical proof that such an algorithm does not exist was found.

11.4 Application of Graph Coloring

11.4.1 Scheduling

You may wonder why we should be interested in graph colorings at all. Is this merely an artistic challenge? Not at all. In fact, this problem has many important applications, two of which we explore in this section.

Designing a time table for high school, college, or any big educational institute where students have individual preferences, and the system has to accommodate many of these preferences, is a very hard task. Do not trust us – talk to the school or college person in charge of this task. For the sake of simplicity, consider some arts high school, and let us restrict ourselves to the scheduling of classes on Monday alone. Furthermore, suppose that due to teachers' availabilities, the following seven classes are to be taught on Monday: music, art, English, math, science, economics, and accounting. The duration of each class is the same – two hours each. Classes start at 8am, so the following time slots are utilized: 8–10, 10–12, 12–2, etc. We can easily find a scheduling that will meet this "no overlap" criterion by having each course set for its own time slot. However, in this way we will have seven sessions of two hours each, and an overall school day of 14 hours. Obviously, this is unacceptable. So let us add one more goal – we want the total number of time slots to be minimized, and not exceed three time slots (8–10, 10–12, 12–14).

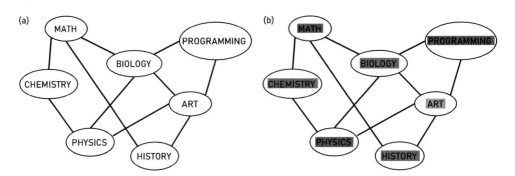

Figure 11.5. (a) A graph representing a scheduling constraints. Two courses whose nodes are connected cannot be given at the same time slot. (b) The graph colored in three colors. Red, blue and green represent the time slots 8–10, 10–12, 12–14 correspondingly

The basic observation is that if the sets of students taking two or more courses are distinct, we can schedule them in the same time slot. Furthermore, we can represent these constraints using graphs. We construct an undirected graph with seven nodes, each corresponding to one of the subjects above. We put an undirected edge between two nodes ("subjects") if they should not be scheduled in the same time slot. An example for a set of constraints is depicted in Figure 11.5(a).

How does all this relate to graph coloring? Let us represent the time slots 8–10, 10–12, and 12–14 with the colors red, blue, and green, correspondingly. Since we aimed for not exceeding three time slots, what we are actually looking for is a legal coloring of the constraints graph using at most three colors! A possible coloring appears in Figure 11.5(b). As we succeeded in coloring the graph using just three colors, this implies that three time slots are sufficient, and on Monday classes can end by 2pm, making both the students and the teachers happy.

We comment that our success in finding an optimal solution to the coloring of this specific graph is mainly due to the fact that this is a rather small graph. As you now know, coloring (and thus scheduling) is a problem we currently have no efficient solutions for.

What we saw here is that certain scheduling problems can be *reduced* to graph coloring, so that a solution to the coloring problem immediately translates to a solution to the corresponding scheduling problem. This does not imply that scheduling or coloring are computationally feasible, but that their computational hardness is tightly related. If we miraculously find an efficient way to solve one, this will imply an efficient way to solve the other.

11.4.2 Cell Phone Frequencies

A cellular phone network employs a large number of fixed location antennas to cover much of the inhabited areas worldwide. Each such antenna serves a relatively small area, called a cell. The range of such antennas is typically larger than the cell itself, and thus the same point in a city may be covered by more than a single antenna. To avoid interference and provide good service, if the coverage area of two or more antennas overlap, they are typically assigned

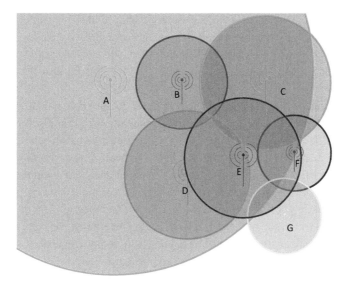

Figure 11.6. Seven antennas with overlapping coverage areas

different frequencies. Frequencies for cell phone communication are a very limited resource, and are often sold at auctions for substantial amounts of money. So we are faced with a different constraints problem: avoid conflicts while using the minimum number of frequencies.

A simple instance of this frequencies allocation problem is depicted in Figure 11.6. We have seven antennas, termed A to G. The coverage area of each antenna has a circular shape, but the areas are not all the same. Areas covered by more than a single antenna are shaded darker. Suppose leasing one frequency costs one million dollars per year. The cost of assigning a different frequency per antenna would thus be 7 million dollars per year. Can we do better?

Here, again, we can translate the frequencies assignment problem into a graph coloring problem. The nodes of our graph will be the antennas, and we put an undirected edge between two antennas with overlapping coverage. Each frequency corresponds to a color in this graph. Note that this abstraction ignores many physical and other aspects of the problem, which are irrelevant to solving the frequencies allocation problem.

Exercise 2

Draw the corresponding graph. Can you color it using: Three colors? Four colors? Five colors? Six colors?

11.5 Verifying a Given Coloring and the P vs. NP Problem

Let us now look at a different angle of coloring. Suppose someone gives you a coloring of some graph G and claims it is a legal k-coloring of G. Such a claim can be efficiently **verified**. Counting the number of colors, you verify it is not larger than k. And then going over all edges, you verify the two nodes at the ends are assigned different colors by the given coloring. If both tests are passed successfully, you are convinced by the given coloring that the graph is indeed k-colorable. If this verification failed, you can reject this specific claim for the specific coloring. Such rejection has no bearing on the correctness of the claim that the graph is k-colorable.

The conclusion is this: despite the fact that we do not know how to efficiently *find* a legal k-coloring of G by ourselves, we are able to efficiently *verify* for any $k > 2$ that a given coloring is a legal k-coloring of G. This is an example of a common scenario in computer science: there are many important problems where **verification** is easy, while it is unknown how to construct an efficient solution. We emphasize that not all hard computational problems have the property that they are efficiently verifiable. For example, we do not know how to efficiently verify that a given graph G on n nodes is *not* k-colorable.

A major question for all these problems having an efficient verifier is whether or not they can be solved efficiently. This problem, known in computer science as the **P vs. NP** problem, has attracted the interest of many researchers since the early 1970s. P is defined as the class of problems that have an efficient (polynomial time) solution. NP is the class of problems for which one can verify a given solution. So the P vs. NP problem translates to: can we efficiently solve any problem that can be efficiently verified? The Clay institute has offered a monetary prize of USD 1,000,000 for the first correct solution of the P vs. NP problem.

11.6 Concluding Remarks and Reflections

We touched upon several important notions in this chapter. One is **abstraction**. In various problems, we are presented with inputs that contain many details, which turn out to be superficial to the task we are asked to solve. In the map coloring problem, maps of physical entities contain many details, depending on the type of maps considered: the exact contours of the borders between entities; the topography; details on cities and villages; land coverage by forests, trees, valleys, rivers, springs, lakes, dams, mines, roads, railroad, industrial areas, etc. All these are important in many settings, but are inconsequential for maps coloring. The only thing that matters is which entities border each other. The graph that corresponds to the map conveys exactly this information, and nothing else. It abstracts away all superficial details, retaining only the essential information.

Problems such as the coloring and scheduling we saw have the property that their solutions are efficiently verifiable. If one presents a proposed solution, it is possible to efficiently verify that this is indeed a solution (or refute this if, for example, a proposed coloring is not legal).

Remember the Eulerian and Hamiltonian paths problem from chapter 5? The Eulerian path problem can be solved efficiently, as we saw, thus it is in P. The Hamiltonian path problem can be verified efficiently (given a path that claims to visit every node in a graph exactly once, we can easily check this to be true or false), thus is it in NP. Today, we have no efficient solution to the last problem, and so we do not know whether it is also in P or not. However, it is widely believed that it is not in P. If this is indeed the case, we will never have an efficient solution to the Hamiltonian paths problem.

The fact that a problem belongs to this seemingly hard class of problems probably means that it would be hard to solve it for large inputs. However, often there are approaches to efficiently solve such problems in practice. In many cases, smart exponential algorithms work well for moderate size inputs. Heuristics can give good outputs for many, but probably not all, moderate size instances. Many problems have approximate solutions, which are often acceptable (e.g., an algorithm guaranteed to produce a 5 coloring of a graph, in case it has a 3 coloring).

Index